70037437

519. PLE

advancing learning, changing lives

D0809927

REVISE M1

Edexcel AS and A Level Modular Mathematics

Mechanics 1

John Hebborn
Jean Littlewood

Published by Pearson Education Limited, a company incorporated in England and Wales, having its registered office at Edinburgh Gate, Harlow, Essex, CM20 2JE. Registered company number: 872828

Edexcel is a registered trademark of Edexcel Limited

Text © John Hebborn and Jean Littlewood 2001, 2009

First published 2001 under the title *Heinemann Modular Mathematics for Edexcel AS and A level: Revise for Mechanics 1*

12 11 10 09
10 9 8 7 6 5 4 3 2 1

British Library Cataloguing in Publication Data is available from the British Library on request.

ISBN 978 0 435519 32 2

Edited by Jim Newall
Typeset by Tech-Set Ltd
Illustrated by Tech-Set Ltd
Cover design by Christopher Howson
Picture research by Chrissie Martin
Cover photo/illustration © Edexcel
Printed in the UK by Scotprint

Acknowledgements
Every effort has been made to contact copyright holders of material reproduced in this book.
Any omissions will be rectified in subsequent printings if notice is given to the publishers.

About this book

This book is designed to help you get your best possible grade in your Mechanics 1 examination. The authors were Chief and Principal examiners, and have a good understanding of Edexcel's requirements.

Revise for Mechanics 1 covers the key topics that are tested in the Mechanics 1 examination paper. You can use this book to help you revise at the end of your course, or you can use it throughout your course alongside the course textbook, *Edexcel AS and A-level Modular Mathematics Mechanics 1*, which provides complete coverage of the specification.

Helping you prepare for your examination

To help you prepare, each topic offers you:

- **What you should know** – a summary of the ideas you need to know and be able to use.

- **Test yourself questions** – help you see where you need extra revision and practice. If you do need extra help they show you where to look in the *Edexcel AS and A-level Modular Mathematics Mechanics 1* textbook.

- **Worked examples and examination questions** – help you understand and remember important methods, and show you how to set out your answers clearly.

- **Revision exercises** – help you practise using these important methods to solve problems. Examination-level questions are included so you can be sure that you are reaching the right standard, and answers are given at the back of the book so that you can assess your progress.

Examination practice and advice on revising

Examination style paper – this paper at the end of the book provides a set of questions of examination standard. It gives you an opportunity to practise taking a complete examination before you meet the real thing. The answers are given at the back of the book.

How to revise – for advice on revising before the examination, read the How to revise section on the next page.

How to revise using this book

Making the best use of your revision time

The topics in this book have been arranged in a logical sequence so you can work your way through them from beginning to end. However, **how** you work on them depends on how much time there is between now and your examination.

If you have plenty of time before the examination then you can **work through each topic in turn**, covering the key points and worked examples before doing the revision exercises and test yourself questions.

If you are short of time then you can **work through the Test yourself sections** first, to help you see which topics you need to do further work on.

However much time you have to revise, make sure you break your revision into short blocks of about 40 minutes, separated by five- or ten-minute breaks. Nobody can study effectively for hours without a break.

Using the Test yourself sections

Each Test yourself section provides a set of key questions. Try each question.

- If you can do it and get the correct answer then move on to the next topic. Come back to this topic later to consolidate your knowledge and understanding by working through the what you should know section, worked examples and revision exercises.

- If you cannot do the question, or get an incorrect answer or part answer, then work through the what you should know section, worked examples and revision exercises before trying the Test yourself questions again. If you need more help, the cross-references beside each Test yourself question show you where to find relevant information in the *Edexcel AS and A-level Modular Mathematics Mechanics 1* textbook.

Reviewing what you should know

Most of what you should know are straightforward ideas that you can learn: try to understand each one. Imagine explaining each idea to a friend in your own words, and say it out loud as you do so. This is a better way of making the ideas stick than just reading them silently from the page.

As you work through the book, remember to go back over what you should know sections from earlier topics at least once a week. This will help you to remember them in the examination.

Mathematical models in mechanics

1

What you should know

1 The following terminology is used in modelling in mechanics:

(i) A **particle** is a body whose dimensions are so small compared with the other lengths involved that its position in space can be represented by a single point. For example, in considering the motion of the Earth relative to the Sun you may represent the Earth and the Sun by particles.

(ii) A **bead** is a particle which is assumed to have a hole drilled through it so that it may be threaded onto a string or wire. For example the plastic animal threaded on a wire as part of a child's rattle may be modelled as a bead.

(iii) A **lamina** is a flat object whose thickness is small compared with its width and length. For example, a piece of card, a sheet of paper or a thin metal sheet may be represented by laminae (plural of lamina).

(iv) A **uniform lamina** is one in which equal areas of the lamina have equal masses. This is clearly the case when the whole of the lamina is made of the same material and has constant thickness.

(v) A **rigid body** is an object made up of particles, all of which remain at the same fixed distances from one another whether the object is at rest or in motion. For example, a rigid beam is assumed to keep its shape when acted on by forces. When a billiard ball strikes the edge of a snooker table you may assume it does not change its shape, so it can be represented by a rigid body.

(vi) A **wire** is a rigid body in the form of a thin thread of metal. The wire may be smooth or rough. We often consider beads threaded on wires.

(vii) A **rod** is an object all of whose mass is concentrated along a line. It is assumed to have length only, and its width and breadth are neglected. A broom pole may, for example, be modelled by a rod.

(viii) A **uniform rod** is one in which equal lengths have equal masses.

(ix) A rod which is not uniform is said to be **non-uniform**. In this case equal lengths do not have equal masses. An example of a non-uniform rod is a baton made in two sections which are different woods.

(x) A **light object** is one whose mass is so small compared with other masses being considered that the mass may be considered to be zero. A **light string** is one example. If an object is suspended by a light string, the mass of the string may be ignored. A **light rod** is another example. If two particles are joined by a light rod these particles remain the same distance apart. The mass of the rod may be neglected.

(xi) An **inextensible string** or **inelastic string** is a string whose length remains the same whether motion is taking place or not. While all real strings are elastic to some extent, in many problems the extension is so small compared to the other lengths under consideration that it may be ignored.

(xii) A **smooth surface** is one which offers so little frictional resistance to the motion of a body sliding across it that the friction may be ignored. A sheet of ice is an example of a smooth surface commonly found in the real world.

(xiii) A surface which is not smooth is said to be **rough**. Frictional forces have to be taken into account when a body moves on such a surface. An example of a rough surface is the surface of a hard tennis court.

(xiv) A **smooth pulley** is one with no friction in its bearings.

(xv) A **peg** is a pin or support from which a body may be hung or on which a body may rest. There is only one point of contact between the peg and the body in either case (see figures). The peg may be either smooth or rough. In the first case there is only a contact force. In the second case frictional forces will also act.

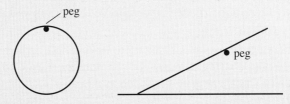

(xvi) A **plane surface** is a completely flat surface. Walls, floors and tables are usually modelled by plane surfaces.

(xvii) The **Earth's surface** is usually modelled by a horizontal plane surface. (You may also model slopes by inclined planes.)

(xviii) A given mathematical model may not adequately describe a physical situation. In such a case the model must be **refined** by taking into account factors originally ignored.

2 Mathematical modelling involves:

(i) **translating** a real world problem into a mathematical problem or model;

(ii) **solving** the mathematical problem;

(iii) **interpreting** the solution in terms of the real world.

The process can be shown as a diagram:

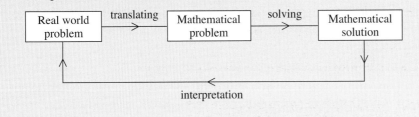

Test yourself

This chapter is short and yet it contains important ideas that lie behind much of mechanics. If you are confident you can answer the two questions in Revision exercise 1 then do so and check your answer against those at the back of the book. If you have any doubts read through this section before attempting them.

What to review

If your answer is incorrect:

Review Edexcel Book M1 pages 1–3

Revise for M1 pages 1–3

Example 1

Suggest a simple mathematical model for the situation:

　　'An ice hockey player strikes a puck on an ice rink.'

> A particle receives an impulse and moves on a smooth horizontal plane. The puck is modelled by a particle and the ice rink by a smooth horizontal plane.

Example 2

'A tennis ball is hit by a player. Determine its subsequent motion.'

This situation was modelled by:

　'A particle moves freely under a constant vertically downward force. Obtain its trajectory.'

Comment on this model and suggest how it could be refined.

> The model is a reasonable initial model since a tennis ball is fairly small compared with the distances involved and g is not likely to vary much on this path. In addition, for most tennis players the speed of the ball is likely to be such that air resistance is small.
>
> The model could be refined by taking into account the variation of g and the air resistance. The size of the ball and the possibility of spin could also be included.

Revision exercise 1

1　Give examples where it is appropriate to model a body by a particle.

2　A small box is dragged across the playground by a rope inclined at 30° to the horizontal. Suggest an initial model for this situation and list some refinements that could be made.

Kinematics of a particle moving in a straight line

2

What you should know

1 For a body to be modelled as a **particle** its dimensions must be so small that its position can be represented by a point.

2 For a particle moving with **constant acceleration**:

$$v = u + at$$

$$s = \left(\frac{u + v}{2}\right)t$$

$$s = ut + \tfrac{1}{2}at^2$$

$$v^2 = u^2 + 2as$$

where a is the *constant* acceleration, u is the initial velocity, v is the final velocity and s the distance travelled in time t.

3 A negative acceleration is called a **retardation** or **deceleration**.

4 Provided the effect of air resistance can be ignored and no forces other than gravity are acting, a particle moving vertically up or down will have a constant acceleration g. This acceleration acts downwards and has a numerical value of approximately $9.8 \, \mathrm{m \, s^{-2}}$. Answers obtained from using this value should be given to 2 significant figures.

5 A particle which is *dropped* or *starts from rest* will have initial speed zero.

6 A particle *moving freely under gravity* is at its highest point when its velocity is zero.

7 On a speed–time graph time is always plotted along the **horizontal** axis.

8 For a particle moving with constant acceleration the speed–time graph will be a **straight line**.

9 The **gradient** of the speed–time graph is the acceleration of the particle.

10 The **area** under the speed–time graph is equal to the distance travelled by the particle.

11 Unless you are told to draw an accurate graph a *sketch* on ordinary lined paper is all that is required.

Test yourself

1 A car is moving along a straight road with uniform acceleration. The car passes a check-point A with a speed of $12\,\mathrm{m\,s^{-1}}$ and another check-point C with a speed of $32\,\mathrm{m\,s^{-1}}$. The distance between A and C is 1100 m.

(a) Find the time, in seconds, taken by the car to move from A to C.

Given that B is the mid-point of AC:

(b) find, in $\mathrm{m\,s^{-1}}$ to 1 decimal place, the speed with which the car passes B. [E]

Review Edexcel Book M1 pages 5–15

Revise for M1 pages 6–14

2 A ball is thrown vertically upwards at $4.9\,\mathrm{m\,s^{-1}}$ from a window which is 5 m above horizontal ground. By modelling the ball as a particle moving freely under gravity, find:

(a) the greatest height above the ground attained by the ball

(b) the time, in seconds to 2 significant figures, taken by the ball to reach the ground.

(c) State two assumptions you have made about the ball and the forces acting on it during its motion.

Review Edexcel Book M1 pages 17–22

Revise for M1 pages 6–14

3 A car is travelling along a straight motorway at a constant speed $V\,\mathrm{m\,s^{-1}}$. Ten seconds after passing a speed-limit sign, the driver brakes and the car decelerates uniformly for 5 seconds, reducing its speed to $30\,\mathrm{m\,s^{-1}}$.

(a) Sketch a speed–time graph to illustrate this information.

Given that the car covers a distance of 600 m in the 15 second period, find:

(b) the value of V

(c) the deceleration of the car. [E]

Review Edexcel Book M1 pages 24–27

Revise for M1 pages 6–14

4 Two cars A and B are travelling along the same straight road. A is travelling at a constant speed of $30\,\mathrm{m\,s^{-1}}$ when it overtakes B which is travelling at a constant speed of $25\,\mathrm{m\,s^{-1}}$. T seconds later A starts to decelerate uniformly coming to rest 1000 m from the point where A overtook B. B maintained the constant speed of $25\,\mathrm{m\,s^{-1}}$ for 30 s and then decelerated uniformly. A and B both come to rest at the same point at the same instant.

(a) Sketch the speed–time graphs of the two cars on the same diagram.

(b) Calculate the time for which B was decelerating.

(c) Calculate the value of T.

Review Edexcel Book M1 page 28

Revise for M1 pages 6–14

The constant acceleration equations are used in many types of questions in the M1 examination, not just the type that appear in this section. They are not given in the formula book; it is therefore important that you know them well enough to be able to quote them accurately without hesitation.

For any moving particle a graph can be drawn of its speed against the time. For M1, only graphs of particles moving with constant acceleration need be considered.

Example 1

A vertical cliff is 98 m high. At time $t = 0$ a particle P is projected vertically upwards from the foot of the cliff with a speed of $45\,\text{m}\,\text{s}^{-1}$. Three seconds later a particle Q is projected downwards from the top of the cliff with a speed of $5\,\text{m}\,\text{s}^{-1}$. Given that P and Q are moving in the same vertical line, find:

Accelerates uniformly means having constant acceleration.

(a) the value of t, in seconds, when P and Q collide

(b) the distance, in metres, below the top of the cliff of the point of collision.

P and Q collide when they are at the *same point* at the *same time*. Let the point of collision be x m below the top of the cliff.

Consider each particle in turn to form two equations to find the two required quantities.

Work with upwards as positive here as that introduces fewer negative signs.

(a) For P, from projection to collision, the known quantities are $a = -9.8\,\text{m}\,\text{s}^{-2}$, $u = 45\,\text{m}\,\text{s}^{-1}$, $s = (98 - x)$ m.

Using **2**, $\qquad s = ut + \frac{1}{2}at^2$

gives: $\qquad 98 - x = 45t - \frac{1}{2} \times 9.8t^2$

Remember the acceleration due to gravity is always downwards.

$\qquad\qquad 98 - x = 45t - 4.9t^2 \qquad\qquad (1)$

For Q, from projection to collision, the known quantities are $A = 9.8\,\text{m}\,\text{s}^{-2}$, $U = 5\,\text{m}\,\text{s}^{-1}$, $S = x$ m.

Take downwards as positive here to introduce fewer negative signs.

Q is thrown 3 s after P so the time T for Q is given by $T = t - 3$

Using **2**, $\quad S = UT + \frac{1}{2}AT^2$

gives: $\qquad x = 5(t - 3) + \frac{1}{2} \times 9.8(t - 3)^2$

$\qquad\qquad x = 5(t - 3) + 4.9(t - 3)^2 \qquad\qquad (2)$

Adding equations (1) and (2) will eliminate x:

$\qquad\qquad 98 = 45t + 5(t - 3) - 4.9t^2 + 4.9(t - 3)^2$

$\qquad\qquad 98 = 45t + 5t - 15 - 4.9t^2$

$\qquad\qquad\qquad + 4.9(t^2 - 6t + 9)$

$\qquad\qquad 98 = 50t - 15 - 4.9 \times 6t + 4.9 \times 9$

$$98 + 15 - 4.9 \times 9 = t(50 - 4.9 \times 6)$$

$$t = \frac{113 - 4.9 \times 9}{50 - 4.9 \times 6}$$

$$t = 3.344\ldots$$

The particles collide at time 3.3 s. •

Give answers to 2 significant figures as $g = 9.8\,\text{m s}^{-2}$ has been used.

(b) Equation (2) gives the easier solution for x.
Substituting the value of t in equation (2) gives:

$$x = 5(3.344\ldots - 3) + 4.9(3.344\ldots - 3)^2$$

$$= 2.305\ldots$$

The particles collide 2.3 m below the top of the cliff.

Where possible use stored answers from your calculator in further calculations. Writing 3.344… indicates to the examiner that you are doing this.

2

Worked examination question 1 [E]

A particle moves in a straight line ABC starting from rest at A. The particle accelerates uniformly at $2\,\text{m s}^{-2}$ from A to B, where $AB = 9$ m. From B to C the particle retards uniformly for 2 seconds and comes to rest at C. Calculate:

Accelerates uniformly means having constant acceleration.

(a) the speed of the particle as it passes through B

(b) the distance AC.

(a) The known quantities for the motion from A to B are
$s = 9\,\text{m}, a = 2\,\text{m s}^{-2}, u = 0\,\text{m s}^{-1}$.

Using **2**, $\quad v^2 = u^2 + 2as$

gives: $\qquad v^2 = 0 + 2 \times 2 \times 9$

So: $\qquad v = 6$

The particle passes through B with speed $6\,\text{m s}^{-1}$.

(b) The acceleration changes as the particle passes through B. So the distance AC must be obtained by calculating the distance BC and then adding that to the distance AB.

For the motion from B to C the known quantities are
$y = 6\,\text{m s}^{-1}, v = 0\,\text{m s}^{-1}, t = 2\,\text{s}$.

Using **2**, $\quad s = \left(\frac{u + v}{2}\right)t$

gives: $\qquad s = \left(\frac{6 + 0}{2}\right) \times 2$

so: $\qquad s = 6$

And the total distance from A to C is $(6 + 9)\,\text{m} = 15\,\text{m}$.

Worked examination question 2 [E]

A particle P, starting from rest at A, moves in a straight line $ABCD$. It accelerates uniformly at 3 m s^{-2} from A to B. From B to C it travels at a constant velocity, and from C to D retards uniformly at 2 m s^{-2}, coming to rest at D. Given that $AB = 6 \text{ m}$ and that the total time that P is in motion is 12 s, find the distance BC.

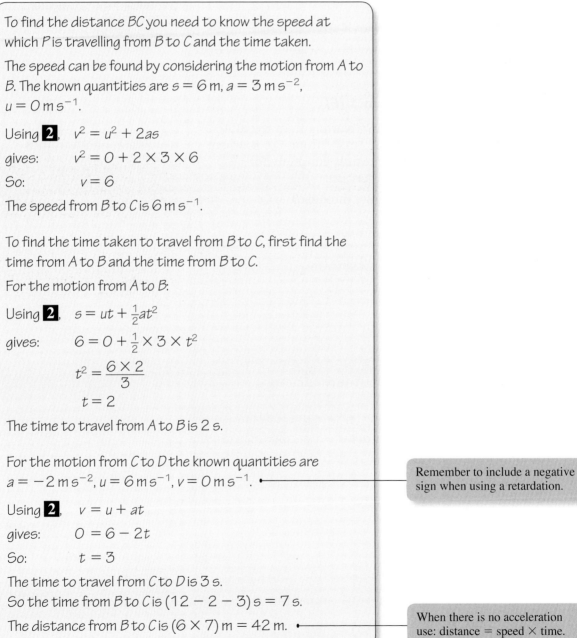

To find the distance BC you need to know the speed at which P is travelling from B to C and the time taken.

The speed can be found by considering the motion from A to B. The known quantities are $s = 6 \text{ m}$, $a = 3 \text{ m s}^{-2}$, $u = 0 \text{ m s}^{-1}$.

Using **2**, $v^2 = u^2 + 2as$

gives: $v^2 = 0 + 2 \times 3 \times 6$

So: $v = 6$

The speed from B to C is 6 m s^{-1}.

To find the time taken to travel from B to C, first find the time from A to B and the time from B to C.

For the motion from A to B:

Using **2**, $s = ut + \frac{1}{2}at^2$

gives: $6 = 0 + \frac{1}{2} \times 3 \times t^2$

            $t^2 = \dfrac{6 \times 2}{3}$

            $t = 2$

The time to travel from A to B is 2 s.

For the motion from C to D the known quantities are $a = -2 \text{ m s}^{-2}$, $u = 6 \text{ m s}^{-1}$, $v = 0 \text{ m s}^{-1}$.

Using **2**, $v = u + at$

gives: $0 = 6 - 2t$

So: $t = 3$

The time to travel from C to D is 3 s.

So the time from B to C is $(12 - 2 - 3) \text{ s} = 7 \text{ s}$.

The distance from B to C is $(6 \times 7) \text{ m} = 42 \text{ m}$.

Remember to include a negative sign when using a retardation.

When there is no acceleration use: distance = speed × time.

Worked examination question 3 [E]

The top of a cliff is 160 m above the beach. A stone is dropped from the top of the cliff and falls freely from rest. Calculate:

(a) the speed with which the stone hits the beach

(b) the time which elapses from the moment when the stone is released until the stone hits the beach.

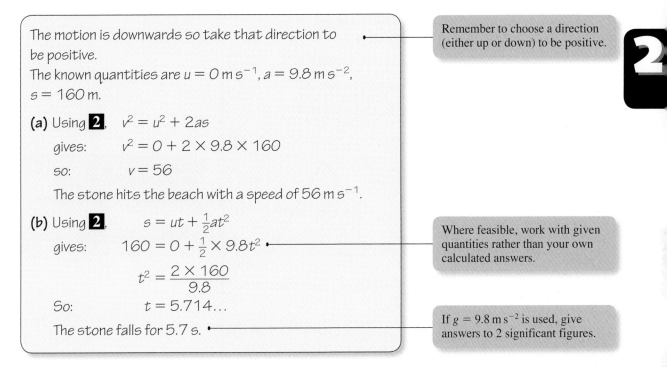

The motion is downwards so take that direction to be positive.

The known quantities are $u = 0 \, \text{m s}^{-1}$, $a = 9.8 \, \text{m s}^{-2}$, $s = 160 \, \text{m}$.

(a) Using **2**, $v^2 = u^2 + 2as$

gives: $v^2 = 0 + 2 \times 9.8 \times 160$

so: $v = 56$

The stone hits the beach with a speed of $56 \, \text{m s}^{-1}$.

(b) Using **2**, $s = ut + \frac{1}{2}at^2$

gives: $160 = 0 + \frac{1}{2} \times 9.8t^2$

$$t^2 = \frac{2 \times 160}{9.8}$$

So: $t = 5.714\ldots$

The stone falls for $5.7 \, \text{s}$.

Remember to choose a direction (either up or down) to be positive.

Where feasible, work with given quantities rather than your own calculated answers.

If $g = 9.8 \, \text{m s}^{-2}$ is used, give answers to 2 significant figures.

Worked examination question 4 [E]

A rocket stands on horizontal ground and is fired vertically from rest with a resultant constant acceleration of $5 \, \text{m s}^{-2}$. After 70 s the motors of the rocket cut out and the rocket moves under the force of gravity only. Sketch a speed–time graph of the motion of the rocket to the highest point of its flight.

Hence, or otherwise, calculate:

(a) the speed of the rocket when the motors cut out

(b) the greatest height reached by the rocket.

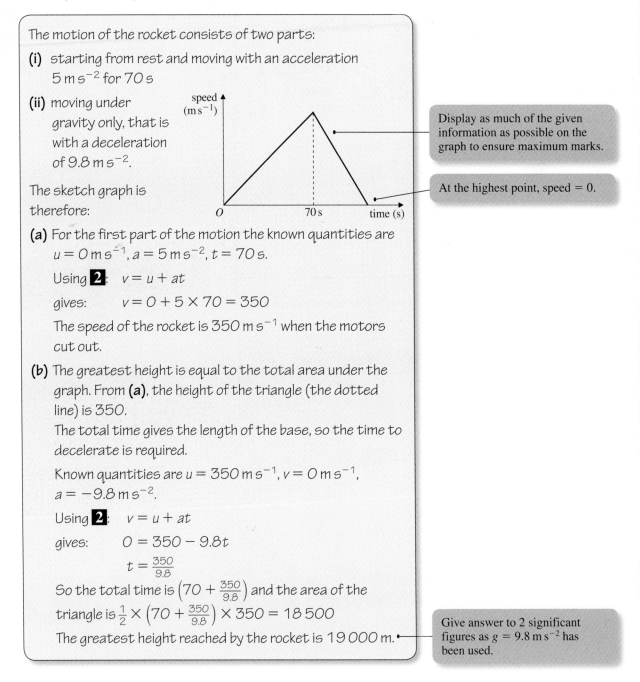

The motion of the rocket consists of two parts:

(i) starting from rest and moving with an acceleration $5 \, \text{m s}^{-2}$ for 70 s

(ii) moving under gravity only, that is with a deceleration of $9.8 \, \text{m s}^{-2}$.

The sketch graph is therefore:

> Display as much of the given information as possible on the graph to ensure maximum marks.

> At the highest point, speed = 0.

(a) For the first part of the motion the known quantities are $u = 0 \, \text{m s}^{-1}$, $a = 5 \, \text{m s}^{-2}$, $t = 70 \, \text{s}$.

Using **2**: $v = u + at$

gives: $v = 0 + 5 \times 70 = 350$

The speed of the rocket is $350 \, \text{m s}^{-1}$ when the motors cut out.

(b) The greatest height is equal to the total area under the graph. From **(a)**, the height of the triangle (the dotted line) is 350.

The total time gives the length of the base, so the time to decelerate is required.

Known quantities are $u = 350 \, \text{m s}^{-1}$, $v = 0 \, \text{m s}^{-1}$, $a = -9.8 \, \text{m s}^{-2}$.

Using **2**: $v = u + at$

gives: $0 = 350 - 9.8t$

            $t = \dfrac{350}{9.8}$

So the total time is $\left(70 + \dfrac{350}{9.8}\right)$ and the area of the triangle is $\dfrac{1}{2} \times \left(70 + \dfrac{350}{9.8}\right) \times 350 = 18\,500$

The greatest height reached by the rocket is $19\,000$ m.

> Give answer to 2 significant figures as $g = 9.8 \, \text{m s}^{-2}$ has been used.

Worked examination question 5 [E]

Two cars A and B are travelling in the same direction along a motorway. They pass a warning sign at the same instant, and subsequently arrive at a toll booth at the same instant.

Car A passes the warning sign at a speed of $24 \, \text{m s}^{-1}$, continues at this speed for one minute, then decelerates uniformly, coming to rest at the toll booth.

Car B passes the warning sign at a speed of $30 \, \text{m s}^{-1}$, continues at this speed for T seconds, then decelerates uniformly, coming to rest at the toll booth.

(a) On the same diagram, sketch the speed–time graph of each car.

The distance from the warning sign to the toll booth is $1.56 \, \text{km}$.

(b) Calculate the length of time, in seconds, for which A was decelerating.

(c) Find the value of T.

2

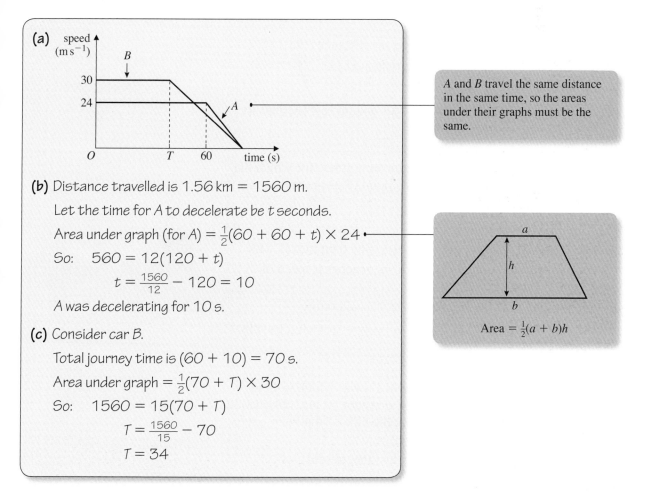

(a)

A and B travel the same distance in the same time, so the areas under their graphs must be the same.

(b) Distance travelled is $1.56 \, \text{km} = 1560 \, \text{m}$.

Let the time for A to decelerate be t seconds.

Area under graph (for A) $= \frac{1}{2}(60 + 60 + t) \times 24$

So: $560 = 12(120 + t)$

$t = \frac{1560}{12} - 120 = 10$

A was decelerating for $10 \, \text{s}$.

Area $= \frac{1}{2}(a + b)h$

(c) Consider car B.

Total journey time is $(60 + 10) = 70 \, \text{s}$.

Area under graph $= \frac{1}{2}(70 + T) \times 30$

So: $1560 = 15(70 + T)$

$T = \frac{1560}{15} - 70$

$T = 34$

Revision exercise 2

1 A car travels a distance of 400 m whilst uniformly decelerating from $40 \, \text{m s}^{-1}$ to $20 \, \text{m s}^{-1}$. Calculate:

(a) the deceleration

(b) the time taken, with this deceleration, to come to rest from a speed of $40 \, \text{m s}^{-1}$.

2 A ball is thrown vertically upwards with a speed of $10 \, \text{m s}^{-1}$ from a point which is 6 m above horizontal ground. By modelling the ball as a particle moving freely under gravity, calculate:

(a) the greatest height above the ground reached by the ball

(b) the speed of the ball, in m s^{-1}, as it strikes the ground

(c) the time, in seconds, for which the ball is in the air.

3 A particle moves with uniform acceleration $2 \, \text{m s}^{-2}$ in a horizontal line PQR. The particle has speed $20 \, \text{m s}^{-1}$ at P and $75 \, \text{m s}^{-1}$ at R. The times taken from P to Q and from Q to R are equal.

(a) Calculate, to the nearest metre, the distance PR.

(b) Calculate the speed of the particle at Q.

4 A stone is thrown vertically upwards and takes 4 seconds to reach its highest point. Calculate:

(a) the speed of projection

(b) the time, in seconds to 2 significant figures, for which the stone is at least 45 m above its point of projection.

5 A cyclist starts from rest and moves along a straight road with constant acceleration $\frac{3}{4} \, \text{m s}^{-2}$ for 10 s. She then moves with constant acceleration $\frac{1}{2} \, \text{m s}^{-2}$ until she reaches a speed of $15 \, \text{m s}^{-1}$. Find:

(a) the total time during which she is accelerating

(b) the total distance, in metres to the nearest metre, travelled whilst accelerating.

6 A box is dropped from a window in a tower which is 40 m above horizontal ground. The box does not hit any obstruction as it falls to the ground. Assuming the box is small enough to be modelled as a particle, find:

(a) the speed with which the box hits the ground

(b) the time taken for the box to hit the ground.

7 A stone is dropped from a top of a cliff. One second later another stone is thrown vertically downwards from the same point with speed $16\,\mathrm{m\,s^{-1}}$. Both stones reach the bottom of the cliff at the same time without hitting any obstructions. Find the height of the cliff.

8 A car is travelling along a straight horizontal road at a constant speed of $20\,\mathrm{m\,s^{-1}}$. It passes a stationary motorcycle and 3 s later the motorcycle follows. The motorcycle accelerates at $2\,\mathrm{m\,s^{-2}}$. Find:

(a) the time, in seconds to 3 s.f. that elapses between the car passing the motorcycle and the motorcycle overtaking the car

(b) the distance, in metres to the nearest metre, travelled by the motorcycle at the instant when it overtakes the car.

9 A train is travelling along a straight track between two stations A and B. It starts from rest at A and accelerates at $2\,\mathrm{m\,s^{-2}}$ until it reaches a speed of $40\,\mathrm{m\,s^{-1}}$. It continues at a constant speed of $40\,\mathrm{m\,s^{-1}}$ for 120s and then decelerates at a constant deceleration of $3\,\mathrm{m\,s^{-2}}$.

(a) Sketch a speed–time graph for the train's journey.

Hence, or otherwise, calculate:

(b) the total time for the journey from A to B

(c) the distance between A and B.

10 A particle starts from rest and accelerates uniformly in a straight line at $f\,\mathrm{m\,s^{-2}}$ for 5 seconds to a speed of $v\,\mathrm{m\,s^{-1}}$.

(a) Write down an expression for v in terms of f.

The particle continues at speed $v\,\mathrm{m\,s^{-1}}$ for 8 seconds. It then decelerates uniformly, coming to rest after a further 2 seconds.

(b) Sketch a speed–time graph for the complete motion.

(c) Given that the total distance travelled is $414\,\mathrm{m}$, find the value of f. [E]

11 A car moving along a straight line accelerates uniformly from rest until it has travelled $x\,\mathrm{m}$. The car then moves for 50 s at constant speed and travels a further $x\,\mathrm{m}$. Finally the car uniformly decelerates and comes to rest after travelling a further $\frac{x}{2}\,\mathrm{m}$. Using a speed–time diagram, or otherwise, calculate the total time for the journey. [E]

12 Two boys A and B are running a 100 m race along a straight track. They leave the starting point at the same time, both starting from rest. A accelerates at $\frac{3}{4}\,\mathrm{m\,s^{-2}}$ to a speed of $4\,\mathrm{m\,s^{-1}}$ and then maintains this speed until he passes the finishing post. B accelerates at $1\,\mathrm{m\,s^{-2}}$ for T seconds and then maintains a constant speed until he passes the finishing post. A and B reach the finishing post at the same time.

(a) On the same diagram, sketch the speed–time graph of each boy.

(b) Find the time taken to run the race.

(c) Find, to 3 significant figures, the value of T.

Dynamics of a particle moving in a straight line or plane

3

What you should know

1 **Newton's laws of motion**
1. A particle will only **accelerate** if it is acted on by a resultant force.
2. The force **F** applied to a particle is **proportional** to the mass m of the particle and the acceleration produced.

   $$\mathbf{F} = m\mathbf{a}$$

3. The forces between two bodies in contact are **equal in magnitude but opposite in direction**.

2 The equation $\mathbf{F} = m\mathbf{a}$ is often called the **equation of motion**.

3 The **weight** of a particle of mass m is given by

   $$\text{weight} = mg$$

4 When two particles are connected by a light inextensible string which passes over a *smooth light* pulley the *tensions* in the parts of the string on either side of the pulley *will be the same*.

> The italics in **4** and **5** show the modelling implications of the situation.

5 Two particles connected by a *light inextensible* string which passes over a smooth pulley will move with the same *acceleration*.

6 The **momentum** of a particle of mass m moving with velocity $\mathbf{v}$ is $m\mathbf{v}$.

7 When a **constant force** $\mathbf{F}$ acts on a particle for time t the **impulse**, $\mathbf{I}$, of the force is equal to $\mathbf{F} \times t$.

8 **Impulse = change in momentum,**

   or $\mathbf{I} = m\mathbf{v} - m\mathbf{u}$

   where $\mathbf{v}$ is the final velocity and $\mathbf{u}$ is the initial velocity.

9   Momentum and impulse are **vector quantities**. Be sure to take account of their directions.

10   For a force in newtons and time in seconds, momentum and impulse are measured in **newton-seconds** (N s).

11   Answers obtained by using $g = 9.8 \, \text{m s}^{-2}$ should be given to 2 significant figures.

12   **Conservation of momentum**
When two particles experience a collision or jerk, where no external forces are involved, momentum is conserved.
$$m_1 u_1 + m_2 u_2 = m_1 v_1 + m_2 v_2$$

13   When two particles P and Q collide the impulse exerted by P on Q is equal in magnitude but opposite in direction to the impulse exerted by Q on P.

14   1 tonne = 1000 kg   1 kN = 1000 N

Test yourself	**What to review**
	If your answer is incorrect:
1 A packing case of mass 20 kg is being pulled across a rough horizontal floor by a rope which is inclined at an angle of 30° to the horizontal. The tension in the rope is 60 N and the packing case has an acceleration of $0.5 \, \text{m s}^{-2}$. Calculate the value of the coefficient of friction between the packing case and the floor.	*Review Edexcel Book M1 pages 50–51* *Revise for M1 pages 18–31*
2 A particle P is projected with speed $12 \, \text{m s}^{-1}$ up a line of greatest slope of a fixed smooth plane inclined at an angle of $\theta°$ to the horizontal. The speed of P after 2 s is $4.8 \, \text{m s}^{-1}$. Find, to 2 significant figures, the value of θ. [E]	*Review Edexcel Book M1 pages 53–55* *Revise for M1 pages 18–31*
3 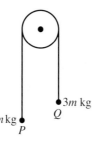	*Review Edexcel Book M1 pages 60–61* *Revise for M1 pages 18–31*

The diagram shows two particles P of mass m kg and Q of mass $3m$ kg which are connected by a light inextensible string which passes over a smooth light fixed pulley. The system is released from rest.

(a) Find the magnitude of the acceleration of the particles.

(b) Find, in terms of m and g, the tension in the string.

4

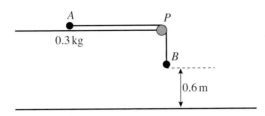

The diagram shows a particle A, of mass $0.3\,\text{kg}$, which rests on a rough horizontal table. A is attached to a particle B, of mass $0.5\,\text{kg}$, by a light inextensible string which passes over a smooth light pulley P which is fixed at the edge of the table. B is hanging $0.6\,\text{m}$ vertically above the horizontal floor. The coefficient of friction between A and the table is 0.4. The system is released from rest with A at a distance of $1\,\text{m}$ from P.

(a) Find, in m s^{-2} to 3 significant figures, the acceleration of the particles.

When B hits the floor it does not rebound.

(b) Find, in m s^{-1} to 3 significant figures, the speed of A as it reaches P.

Review Edexcel Book M1 pages 61–62
Revise for M1 pages 18–31

5 A particle of mass $2\,\text{kg}$ moves horizontally in a straight line with a speed of $3\,\text{m s}^{-1}$. It is given an impulse of magnitude $24\,\text{N s}$ in the opposite direction to its motion. Determine the speed of the particle, in m s^{-1}, immediately after the impulse is applied. [E]

Review Edexcel Book M1 page 67
Revise for M1 pages 18–31

6 A cricket ball, of mass $0.16\,\text{kg}$, is moving horizontally at $19\,\text{m s}^{-1}$ when it hits at right angles a vertical fixed sight-screen. The ball bounces back horizontally from the sight-screen at $11\,\text{m s}^{-1}$.

(a) Find, in N s, the impulse of the force exerted by the ball on the sight-screen.

Given that the ball and the sight-screen are in contact for $0.15\,\text{s}$:

(b) find, in N, the force, assumed constant, exerted by the ball on the sight-screen. [E]

Review Edexcel Book M1 page 67
Revise for M1 pages 18–31

7 A railway truck P of mass 5 tonnes is moving along a straight horizontal track with a speed of $15\,\text{m s}^{-1}$. It collides with a stationary truck Q of mass 7.5 tonnes. The two trucks couple together and move as one truck after the collision.

(a) Find the speed of the trucks immediately after the collision.

(b) Find the magnitude of the impulse exerted on P during the collision.

Review Edexcel Book M1 page 68–71
Revise for M1 pages 18–31

After the collision the trucks are brought to rest by a constant horizontal resistance of magnitude 4.5 kN.

(c) Find the time that elapses between the collision and the trucks coming to rest.

8 A bullet of mass 25 g is fired horizontally from a rifle of mass 6 kg. The bullet leaves the rifle with a speed of $500 \, \text{m s}^{-1}$.

(a) Find, in m s^{-1}, the initial speed of recoil of the rifle.

The rifle recoils a horizontal distance of 2 cm before coming to rest. Assuming that the horizontal force which brings it to rest is constant:

(b) find, in kN to 3 significant figures, the magnitude of the horizontal force.

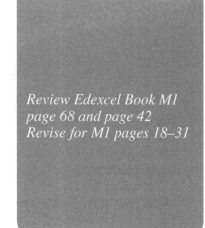

Review Edexcel Book M1 page 68 and page 42
Revise for M1 pages 18–31

Example 1

Two particles P and Q of mass $2m$ and $4m$ respectively are connected by a light inextensible string. P and Q are resting on a smooth horizontal table with the string slack. Q is projected away from P along the table with speed u. At the instant when the string becomes taut P is jerked into motion. Assuming that both particles then move with the same speed in the direction of Q's original motion find:

(a) the speed of P after the string becomes taut

(b) the magnitude of the impulse exerted on P when the string becomes taut.

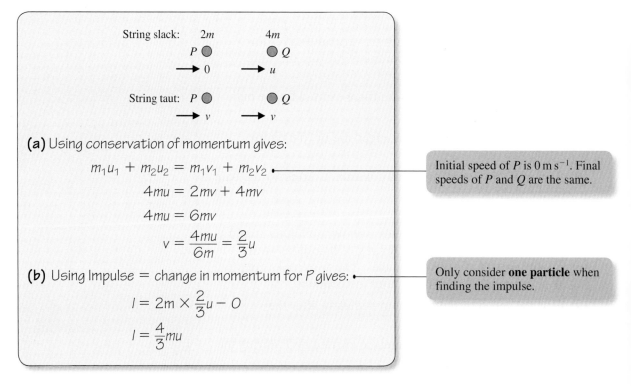

String slack: $2m$ $4m$
         P ●   ● Q
         → 0   → u

String taut: P ●   ● Q
         → v → v

(a) Using conservation of momentum gives:

$$m_1 u_1 + m_2 u_2 = m_1 v_1 + m_2 v_2$$
$$4mu = 2mv + 4mv$$
$$4mu = 6mv$$
$$v = \frac{4mu}{6m} = \frac{2}{3}u$$

Initial speed of P is $0 \, \text{m s}^{-1}$. Final speeds of P and Q are the same.

(b) Using Impulse = change in momentum for P gives:

$$I = 2m \times \frac{2}{3}u - 0$$
$$I = \frac{4}{3}mu$$

*Only consider **one particle** when finding the impulse.*

Example 2

Two particles A and B are moving in the same direction in the same straight line. A has mass $3m$ and speed $5u$. B has mass $6m$ and speed u. The particles collide and coalesce. Find, in terms of u, the speed of the combined particle after the impact.

Coalesce means become a single particle.

Before impact: $\longrightarrow 5u$ $\longrightarrow u$
$A \bullet 3m$ $B \bullet 6m$

After impact: $\bullet 9m$
$\longrightarrow v$

Using conservation of momentum gives:

$$\rightarrow \quad m_1u_1 + m_2u_2 = m_1v_1 + m_2v_2$$
$$3m \times 5u + 6m \times u = 9m \times v$$
$$15u + 6u = 9v$$
$$9v = 21u$$
$$v = \frac{21}{9}u = \frac{7}{3}u$$

The speed of the combined particle is $\frac{7}{3}u$.

3

Example 3

Two particles P and Q have masses λm and m respectively, where λ is a positive constant. P and Q are moving in the same direction in the same straight line. P has speed $5u$ and Q has speed $2u$. P and Q collide. After the collision the particles continue to travel in the same direction and the speed of Q has doubled.

(a) Find, in terms of λ and u, the speed of P after the collision.

(b) Show that $\frac{2}{5} < \lambda < 2$.

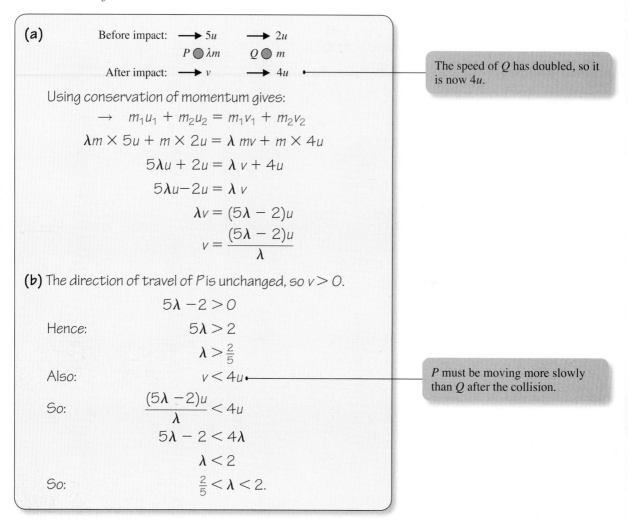

(a)

Before impact: → $5u$ → $2u$

P ● λm Q ● m

After impact: → v → $4u$

> The speed of Q has doubled, so it is now $4u$.

Using conservation of momentum gives:

$$\rightarrow \quad m_1 u_1 + m_2 u_2 = m_1 v_1 + m_2 v_2$$

$$\lambda m \times 5u + m \times 2u = \lambda\, mv + m \times 4u$$

$$5\lambda u + 2u = \lambda v + 4u$$

$$5\lambda u - 2u = \lambda v$$

$$\lambda v = (5\lambda - 2)u$$

$$v = \frac{(5\lambda - 2)u}{\lambda}$$

(b) The direction of travel of P is unchanged, so $v > 0$.

$$5\lambda - 2 > 0$$

Hence:

$$5\lambda > 2$$

$$\lambda > \frac{2}{5}$$

Also:

$$v < 4u$$

> P must be moving more slowly than Q after the collision.

So:

$$\frac{(5\lambda - 2)u}{\lambda} < 4u$$

$$5\lambda - 2 < 4\lambda$$

$$\lambda < 2$$

So:

$$\frac{2}{5} < \lambda < 2.$$

Worked examination question 1 [E]

A constant retarding force of magnitude F newtons reduces the speed of a car, of mass $800\,\text{kg}$ and travelling along a straight horizontal road, from $18\,\text{m s}^{-1}$ to $12\,\text{m s}^{-1}$ in 2.4 seconds. Calculate:

(a) F

(b) the distance covered by the car during this period.

(a) Using: $v = u + at$

with $u = 18\,\text{m s}^{-1}$, $v = 12\,\text{m s}^{-1}$ and $t = 2.4\,\text{s}$, gives:

$$18 = 12 + 2.4a$$

$$a = \frac{6}{2.4}$$

Using $F = ma$, with $m = 800\,\text{kg}$ gives:

$$F = 800 \times \frac{6}{2.4} = 2000$$

(b) Using: $s = \frac{(u + v)}{2}t$

Gives: $s = \frac{(18 + 12)}{2} \times 2.4 = 15 \times 2.4 = 36$

The car travels 36 m.

> The equation of motion will give F but first the acceleration must be found.

> There is no need to work out a value for a when the question does not require it.

Worked examination question 2 [E]

A pebble of mass 0.3 kg slides in a straight line on the surface of a rough horizontal path. Its initial speed is $12.6\,\text{m s}^{-1}$. The coefficient of friction between the pebble and the path is $\frac{3}{7}$.

(a) Find the frictional force retarding the pebble.

(b) Find the total distance covered by the pebble before it comes to rest.

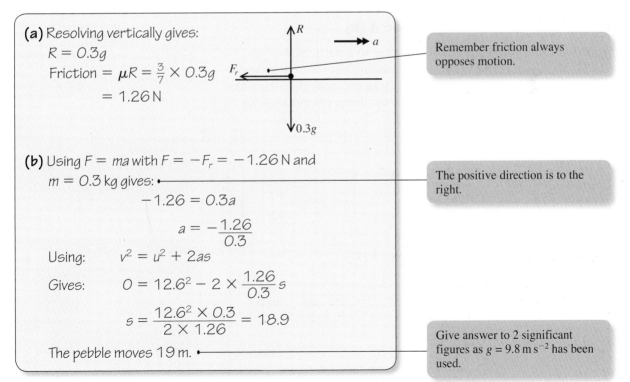

(a) Resolving vertically gives:

$R = 0.3g$

Friction $= \mu R = \frac{3}{7} \times 0.3g$

$= 1.26\,\text{N}$

> Remember friction always opposes motion.

(b) Using $F = ma$ with $F = -F_r = -1.26\,\text{N}$ and $m = 0.3\,\text{kg}$ gives:

$$-1.26 = 0.3a$$

$$a = -\frac{1.26}{0.3}$$

Using: $v^2 = u^2 + 2as$

Gives: $0 = 12.6^2 - 2 \times \frac{1.26}{0.3}s$

$$s = \frac{12.6^2 \times 0.3}{2 \times 1.26} = 18.9$$

The pebble moves 19 m.

> The positive direction is to the right.

> Give answer to 2 significant figures as $g = 9.8\,\text{m s}^{-2}$ has been used.

Worked examination question 3 [E]

A particle, placed on a plane inclined at 30° to the horizontal, moves with constant acceleration of magnitude $1.4\,\mathrm{m\,s^{-2}}$ down a line of greatest slope of the plane. Find the coefficient of friction between the particle and the plane.

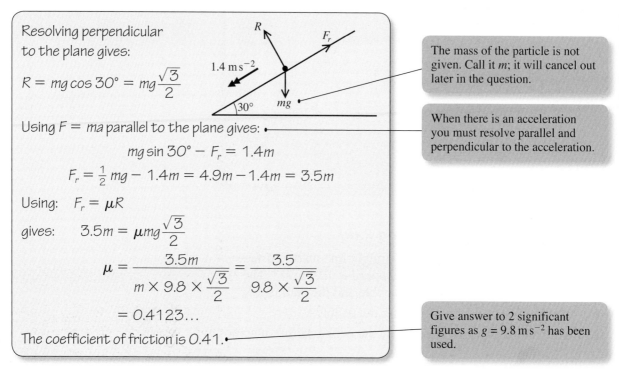

Resolving perpendicular to the plane gives:

$$R = mg\cos 30° = mg\frac{\sqrt{3}}{2}$$

Using $F = ma$ parallel to the plane gives:

$$mg\sin 30° - F_r = 1.4m$$

$$F_r = \tfrac{1}{2}mg - 1.4m = 4.9m - 1.4m = 3.5m$$

Using: $F_r = \mu R$

gives: $3.5m = \mu mg\dfrac{\sqrt{3}}{2}$

$$\mu = \frac{3.5m}{m \times 9.8 \times \dfrac{\sqrt{3}}{2}} = \frac{3.5}{9.8 \times \dfrac{\sqrt{3}}{2}}$$

$$= 0.4123\ldots$$

The coefficient of friction is 0.41.

The mass of the particle is not given. Call it m; it will cancel out later in the question.

When there is an acceleration you must resolve parallel and perpendicular to the acceleration.

Give answer to 2 significant figures as $g = 9.8\,\mathrm{m\,s^{-2}}$ has been used.

Worked examination question 4 [E]

A particle A of mass $0.3\,\mathrm{kg}$ is connected to a particle B of mass $0.4\,\mathrm{kg}$ by a light inextensible string passing over a smooth light fixed pulley. The system is released from rest with the string taut and the hanging parts vertical. Calculate the acceleration with which B descends.

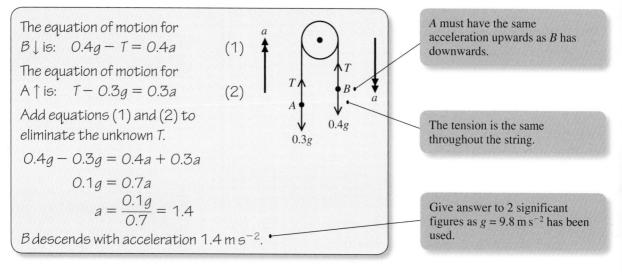

The equation of motion for $B\downarrow$ is: $0.4g - T = 0.4a$ (1)

The equation of motion for $A\uparrow$ is: $T - 0.3g = 0.3a$ (2)

Add equations (1) and (2) to eliminate the unknown T.

$$0.4g - 0.3g = 0.4a + 0.3a$$

$$0.1g = 0.7a$$

$$a = \frac{0.1g}{0.7} = 1.4$$

B descends with acceleration $1.4\,\mathrm{m\,s^{-2}}$.

A must have the same acceleration upwards as B has downwards.

The tension is the same throughout the string.

Give answer to 2 significant figures as $g = 9.8\,\mathrm{m\,s^{-2}}$ has been used.

Worked examination question 5 [E]

A particle A, of mass $0.5\,kg$ is placed on a smooth horizontal table and is connected by a light inextensible string passing over a small smooth pulley at the edge of the table to a particle B, of mass $0.2\,kg$, hanging freely. Each part of the string is taut and perpendicular to the edge of the table. The system is released from rest. Calculate the tension, in N, in the string when the system is in motion.

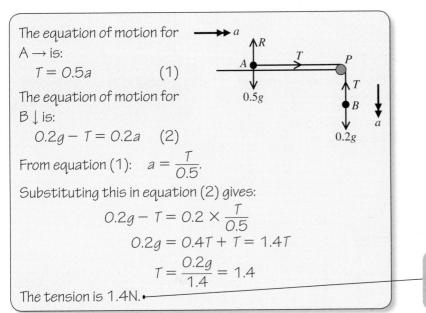

The equation of motion for $A \rightarrow$ is:

$$T = 0.5a \qquad (1)$$

The equation of motion for $B \downarrow$ is:

$$0.2g - T = 0.2a \qquad (2)$$

From equation (1): $a = \dfrac{T}{0.5}$.

Substituting this in equation (2) gives:

$$0.2g - T = 0.2 \times \frac{T}{0.5}$$

$$0.2g = 0.4T + T = 1.4T$$

$$T = \frac{0.2g}{1.4} = 1.4$$

The tension is 1.4N.

Give answer to 2 significant figures as $g = 9.8\,m\,s^{-2}$ has been used.

3

Worked examination question 6 [E]

Two particles, A and B, each of mass m, are connected by a light inextensible string. Particle A is placed on a horizontal table, the string passes over a small smooth light pulley P fixed at an edge of the table and B hangs freely. The horizontal section of the string AP is perpendicular to the edge of the table and of length c. The particles are released from rest with both sections of the string taut and the section PB vertical.

(i) If the table is smooth, find:

   (a) the tension in the string

   (b) the time taken by A to reach the pulley.

(ii) If instead, the table is rough and the coefficient of friction between A and the table is $\frac{1}{4}$, find:

   (c) the tension in the string

   (d) the time taken by A to reach the pulley.

> This could be described as a refined model.

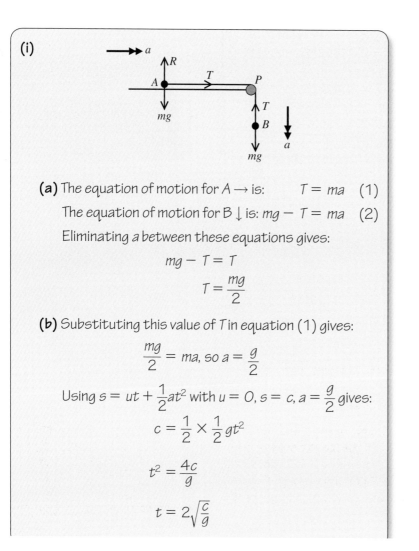

(i)

(a) The equation of motion for $A \rightarrow$ is: $\qquad T = ma$ (1)

The equation of motion for $B \downarrow$ is: $mg - T = ma$ (2)

Eliminating a between these equations gives:

$$mg - T = T$$

$$T = \frac{mg}{2}$$

(b) Substituting this value of T in equation (1) gives:

$$\frac{mg}{2} = ma, \text{ so } a = \frac{g}{2}$$

Using $s = ut + \frac{1}{2}at^2$ with $u = 0, s = c, a = \frac{g}{2}$ gives:

$$c = \frac{1}{2} \times \frac{1}{2}gt^2$$

$$t^2 = \frac{4c}{g}$$

$$t = 2\sqrt{\frac{c}{g}}$$

(ii)

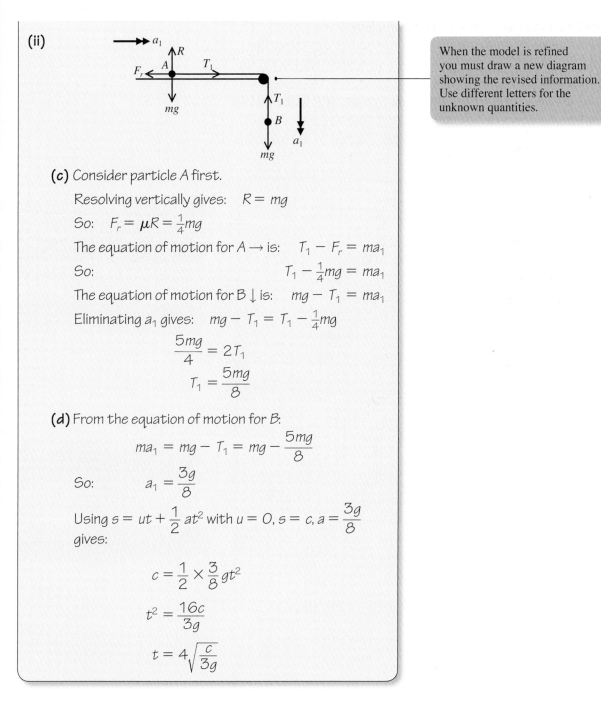

When the model is refined you must draw a new diagram showing the revised information. Use different letters for the unknown quantities.

(c) Consider particle A first.

Resolving vertically gives: $R = mg$

So: $F_r = \mu R = \frac{1}{4}mg$

The equation of motion for $A \rightarrow$ is: $T_1 - F_r = ma_1$

So: $T_1 - \frac{1}{4}mg = ma_1$

The equation of motion for $B \downarrow$ is: $mg - T_1 = ma_1$

Eliminating a_1 gives: $mg - T_1 = T_1 - \frac{1}{4}mg$

$$\frac{5mg}{4} = 2T_1$$

$$T_1 = \frac{5mg}{8}$$

(d) From the equation of motion for B:

$$ma_1 = mg - T_1 = mg - \frac{5mg}{8}$$

So: $a_1 = \frac{3g}{8}$

Using $s = ut + \frac{1}{2}at^2$ with $u = 0, s = c, a = \frac{3g}{8}$ gives:

$$c = \frac{1}{2} \times \frac{3}{8}gt^2$$

$$t^2 = \frac{16c}{3g}$$

$$t = 4\sqrt{\frac{c}{3g}}$$

Worked examination question 7 [E]

The diagram shows two particles P and Q, of mass 1 kg and 0.8 kg respectively, which are connected by a light inextensible string. Particle P rests on a smooth fixed wedge of angle $\alpha°$. The string passes over a smooth light pulley S and Q hangs freely. P, S and Q lie in a vertical plane containing a line of greatest slope of the wedge. The system is in equilibrium.

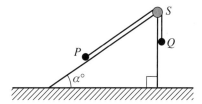

(a) Find $\sin \alpha°$.

The particle Q is now replaced by a particle R of mass 1.4 kg. The system is released from rest with R at a height of 2 m above the horizontal ground. Assuming that P does not strike the pulley, find:

(b) the acceleration, in m s^{-2}, of R

(c) the speed, in m s^{-1} to 2 significant figures, of R at the instant when it reaches the ground

(d) the force, in N to the nearest integer, exerted by the string on the pulley during the motion.

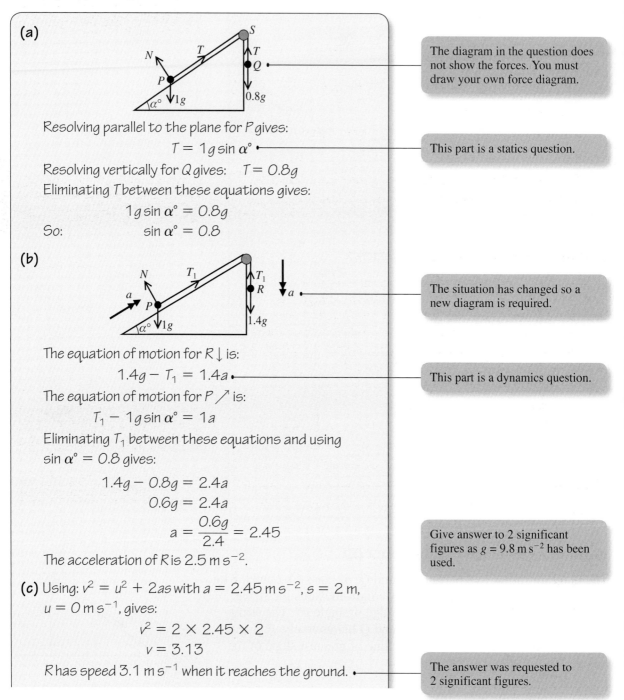

(a)

The diagram in the question does not show the forces. You must draw your own force diagram.

Resolving parallel to the plane for P gives:
$$T = 1g\sin\alpha°$$

This part is a statics question.

Resolving vertically for Q gives: $T = 0.8g$

Eliminating T between these equations gives:
$$1g\sin\alpha° = 0.8g$$

So: $\sin\alpha° = 0.8$

(b)

The situation has changed so a new diagram is required.

The equation of motion for $R\downarrow$ is:
$$1.4g - T_1 = 1.4a$$

This part is a dynamics question.

The equation of motion for $P\nearrow$ is:
$$T_1 - 1g\sin\alpha° = 1a$$

Eliminating T_1 between these equations and using $\sin\alpha° = 0.8$ gives:
$$1.4g - 0.8g = 2.4a$$
$$0.6g = 2.4a$$
$$a = \frac{0.6g}{2.4} = 2.45$$

Give answer to 2 significant figures as $g = 9.8$ m s^{-2} has been used.

The acceleration of R is 2.5 m s^{-2}.

(c) Using: $v^2 = u^2 + 2as$ with $a = 2.45$ m s^{-2}, $s = 2$ m, $u = 0$ m s^{-1}, gives:
$$v^2 = 2 \times 2.45 \times 2$$
$$v = 3.13$$

R has speed 3.1 m s^{-1} when it reaches the ground.

The answer was requested to 2 significant figures.

(d) From the equation of motion for P:

$$T_1 = a + g\sin\alpha° = 2.45 + 0.8 \times 9.8 = 10.29$$

The force on the pulley is the resultant of the two tensions.

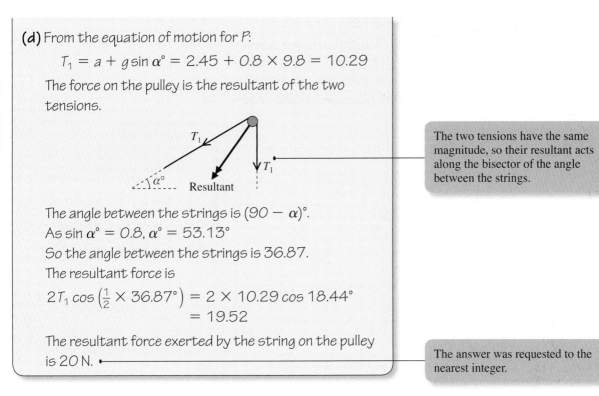

The two tensions have the same magnitude, so their resultant acts along the bisector of the angle between the strings.

The angle between the strings is $(90 - \alpha)°$.

As $\sin\alpha° = 0.8$, $\alpha° = 53.13°$

So the angle between the strings is 36.87.

The resultant force is

$$2T_1 \cos\left(\tfrac{1}{2} \times 36.87°\right) = 2 \times 10.29 \cos 18.44°$$
$$= 19.52$$

The resultant force exerted by the string on the pulley is 20 N.

The answer was requested to the nearest integer.

3

Worked examination question 8 [E]

A tennis ball of mass 0.07 kg is moving horizontally with speed 5 m s^{-1} when it is hit by a racquet and returns straight back horizontally with a speed of 12 m s^{-1}. Calculate, in N s, the magnitude of the impulse of the force of the racquet on the ball. Given that the racquet and the ball are in contact for 0.06 s, find the average force, in N, exerted by the racquet on the ball.

Before impact: → 5 m s^{-1}

0.07 kg ●

After impact: 12 m s^{-1} ←

racquet

A diagram is needed to show the directions for each velocity.

Using **8** ←: $I = mv - mu$

Choose ← as positive direction.

Gives: $I = 0.07 \times 12 - (0.07 \times (-5))$

$I = 0.07 \times 12 + 0.07 \times 5 = 1.19$

The magnitude of the impulse is 1.19 N s.

Using **7**: $I = Ft$

Gives: $1.19 = F \times 0.06$

$$F = \frac{1.19}{0.06} = 19.83...$$

The average force is 19.8 N.

Worked examination question 9 [E]

A ball of mass 0.15 kg falls vertically on to a horizontal concrete floor. The ball strikes the floor with speed 9 m s^{-1} and rebounds at $u \text{ m s}^{-1}$ to reach a height of 2.5 m above the floor. Calculate the value of u and show that the impulse of the force exerted by the ball on the floor is 2.4 N s.

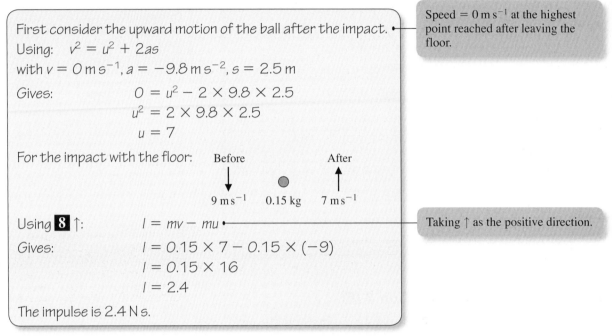

First consider the upward motion of the ball after the impact.

Using: $v^2 = u^2 + 2as$

with $v = 0 \text{ m s}^{-1}, a = -9.8 \text{ m s}^{-2}, s = 2.5 \text{ m}$

Gives: $\qquad 0 = u^2 - 2 \times 9.8 \times 2.5$

$\qquad\qquad u^2 = 2 \times 9.8 \times 2.5$

$\qquad\qquad u = 7$

For the impact with the floor:

Before $\qquad$ After

$9 \text{ m s}^{-1} \qquad 0.15 \text{ kg} \qquad 7 \text{ m s}^{-1}$

Using $\boxed{8}$ ↑: $\qquad I = mv - mu$

Gives: $\qquad I = 0.15 \times 7 - 0.15 \times (-9)$

$\qquad\qquad I = 0.15 \times 16$

$\qquad\qquad I = 2.4$

The impulse is 2.4 N s.

Speed $= 0 \text{ m s}^{-1}$ at the highest point reached after leaving the floor.

Taking ↑ as the positive direction.

Revision exercise 3

1 A ball of mass 10 kg falls vertically through a liquid. The upward force exerted by the liquid on the ball is of magnitude 70 N. Calculate, in m s^{-2}, the acceleration of the ball. [E]

2 A particle moves down a line of greatest slope of a smooth plane inclined at an angle θ to the horizontal. The particle starts from rest and covers 3.5 m in time 2 s. Find the value of $\sin \theta$. [E]

3 A boy is pulling a box of mass 10 kg across rough horizontal ground by means of a rope attached to the side of the box. The coefficient of friction between the box and the ground is 0.6 and the acceleration of the box is 0.2 m s^{-2}. In an initial model the rope is assumed to be horizontal.

(a) State a suitable model for the box.

(b) Calculate, in N, the tension in the rope.

In a refined model, the rope is assumed to be inclined at an angle of 20° to the horizontal. The acceleration is assumed to be unchanged.

(c) Calculate the tension in the rope in this model.

(d) State a physical factor you have ignored in each of your models and give a reason why this factor could be ignored.

4 A particle of mass m starts from rest and moves up a line of greatest slope of a plane inclined at an angle θ to the horizontal, where $\tan \theta = \frac{3}{4}$, under the influence of a force **F** of magnitude $\dfrac{3mg}{2}$ acting parallel to the plane. Given that the coefficient of friction between the particle and the plane is $\frac{1}{2}$, find the acceleration of the particle.

   Calculate, in terms of m, the distance moved by the particle in the first 2 s of its motion.   [E]

5 A particle A, of mass 0.1 kg, is connected to a particle B by a light inextensible string passing over a smooth light fixed pulley. The system is released from rest with the string taut and the hanging parts vertical. Given that B descends with an acceleration $\dfrac{g}{2}$, calculate the mass of B.   [E]

6 A particle P, of mass M, is held at a point A on a rough plane inclined at an angle θ to the horizontal where $\tan \theta = \frac{7}{24}$. The coefficient of friction between P and the plane is $\frac{11}{12}$. A light inextensible string, of length 50 cm, is attached to P and passes over a small smooth pulley O fixed at the top of the inclined plane. To the other end of the string is attached a particle Q, of mass $2M$, which hangs freely and is vertically above a point B on the ground, as shown in the diagram. The distances OA and OB are 32 cm and 43 cm respectively. The particles are released from rest in this position with the string taut and the portion OP parallel to a line of greatest slope of the inclined plane.

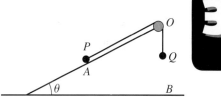

   (a) Show that while the string is taut the acceleration of P up the plane is $\frac{7}{25}g$.

   (b) Find the speed, in m s^{-1} to 2 significant figures, with which Q hits the ground.   [E]

7 The diagram represents two particles A and B of mass 0.2 kg and 0.3 kg respectively, connected by a light inextensible string which passes over a fixed pulley. The particles are released from rest with the string taut and the hanging parts of the string vertical. In the ensuing motion, as there is friction at the pulley, the tension in the part of the string above particle A is T newtons and the tension in the part of the string above particle B is $(T + 0.1)$ newtons.

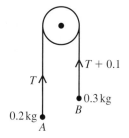

   (a) Write down the equation of motion for each particle.

   (b) Show that the magnitude of the acceleration of each particle is 1.76 m s^{-2}.

   (c) Find the value of T.

At the instant when the particles have a common speed of $5\,\mathrm{m\,s^{-1}}$ the string is cut. Particle B is then $2\,\mathrm{m}$ above the ground.

(d) Calculate, in $\mathrm{m\,s^{-1}}$ to 2 significant figures, the speed with which B hits the ground.

Particle A continues to rise for a further t seconds before coming to instantaneous rest.

(e) Calculate, to 2 decimal places, the value of t. [E]

8 Two small bodies, A and B, of masses $1.2\,\mathrm{kg}$ and $0.9\,\mathrm{kg}$ respectively are attached to the ends of a long, light inextensible string which passes over a smooth vertical light pulley free to turn on a horizontal axis. A bar of mass $0.7\,\mathrm{kg}$ is placed on top of B and the system is released from rest. As B descends it passes through a horizontal ring and the bar is removed instantaneously without altering the speed of B. Calculate:

(a) the acceleration of the system before B reaches the ring

(b) the retardation of A immediately after B passes through the ring

(c) the change in the tension in the string when the bar is removed. [E]

9 A car of mass M kilograms, is pulling a trailer, of mass $\lambda M\,\mathrm{kg}$, along a straight horizontal road. The tow-bar connecting the car and the trailer is horizontal and of negligible mass. The resistive forces acting on the car and trailer are constant and of magnitude $300\,\mathrm{N}$ and $200\,\mathrm{N}$ respectively. At the instant when the car has an acceleration of magnitude $0.3\,\mathrm{m\,s^{-2}}$, the tractive force has magnitude $2000\,\mathrm{N}$. Show that:

$$M(\lambda + 1) = 5000.$$

Given that the tension in the tow-bar is $500\,\mathrm{N}$ at this same instant, find the value of M and the value of λ. [E]

10 A railway engine of mass $5300\,\mathrm{kg}$, moving on horizontal rails at $0.4\,\mathrm{m\,s^{-1}}$, strikes the buffers in a siding and is brought to rest from this speed in $0.2\,\mathrm{s}$. Calculate:

(a) the impulse, in $\mathrm{N\,s}$, of the force exerted by the buffers on the engine in bringing the engine to rest

(b) the magnitude, in N, of this force, assuming it to be constant. [E]

11 A smooth sphere A of mass $0.2\,\mathrm{kg}$ is moving horizontally in a straight line with speed $5\,\mathrm{m\,s^{-1}}$ when it strikes directly a stationary sphere B of equal size and mass $0.1\,\mathrm{kg}$. After the impact A continues in the same direction with speed $2\,\mathrm{m\,s^{-1}}$. Find the speed of B after the impact. [E]

12 A ball of mass 0.4 kg is dropped from a height of 3 m onto horizontal ground. It rebounds to a height of 2 m. Calculate the magnitude of the impulse exerted on the ball by the ground.

13 A bullet is fired horizontally with a speed of 600 m s^{-1} into a block of wood of mass 0.245 kg, resting on a smooth horizontal plane, and becomes embedded in the block. Given that the block begins to move with a speed of 12 m s^{-1}, find, in kg, the mass of the bullet.   [E]

14 Two particles, A and B have masses M and $3M$ respectively. A and B are connected by a light inextensible string. They are placed on a smooth horizontal table with the string taut. A is projected towards B with speed $2u$. After the collision between A and B the speed of A is halved and the direction of travel of A is reversed.

  (a) Find the speed of B after the impact.

  (b) Find the common speed of the particles when the string is taut once more.

  (c) Find the impulse exerted on A at the instant when the string becomes taut.

15 Two particles P and Q of mass $2m$ and m respectively are moving towards each other in the same straight line with speeds $6u$ and $2u$ respectively. The particles collide and after the impact P continues to move in the same direction but the speed of P is halved. Find:

  (a) the speed of Q after the impact

  (b) the magnitude of the impulse exerted by P on Q.

16 The diagram shows two particles P and Q of mass 0.1 kg and 0.2 kg respectively. P and Q are connected by a light inextensible string which passes over a smooth fixed pulley. The system is released from rest with the strings taut and vertical and with Q 0.5 m above a horizontal floor.

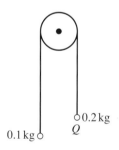

0.1 kg   0.2 kg Q

  (a) Find the acceleration of the particles.

  (b) Find, to 3 significant figures, the speed of Q at the instant when it hits the floor.

When Q hits the floor it does not rebound. Assuming P does not reach the pulley find, to 3 significant figures:

  (c) the speed of Q immediately after it is jerked into motion once more

  (d) the magnitude of the impulse exerted on P as Q is jerked into motion.

17 Two particles P and Q of mass $0.6\,\text{kg}$ and $0.5\,\text{kg}$ respectively are connected by a light string passing over a small smooth fixed pulley. The particles are held at the same height and released from rest with the strings taut and vertical.

(a) Calculate the acceleration of P.

After $1\,\text{s}$ of motion, particle Q picks up a stationary rider of mass $0.3\,\text{kg}$.

(b) Show that the speed of P after the impact is $\frac{9.8}{14}\,\text{m s}^{-1}$.

(c) Calculate the total time that elapses between the instant when P is initially at rest and the next instant P is at rest.

[In this question you may assume that neither particle reaches the pulley during the motion.]   [E]

Statics of a particle

4

What you should know

1 **Force** is a **vector** quantity and is measured in **newtons**.

2 Forces can be added by using the **triangle** or **parallelogram rule**.

3 The component of a force is the product of the magnitude of the force and the cosine of the angle **between** the force and the required direction.

4 The **resultant** of a system of forces is most easily found by using components.

5 A system of forces acting on a particle is in **equilibrium** if their resultant is the **zero vector**.

6 **Friction** acts to oppose relative motion.

7 Until it reaches its limiting value the magnitude of the frictional force is **just sufficient** to prevent relative motion.

8 When sliding occurs the frictional force takes its **limiting value** μR and opposes the relative motion.

9 For a **smooth surface** there is no frictional force ($\mathbf{F} = 0$).

10 When solving problems involving a particle on an inclined plane it is usually preferable to resolve the forces in the directions **parallel** to and **perpendicular** to the plane.

11 A particle of mass m has **weight** mg.

12 Answers which are obtained by using $g = 9.8 \, \mathrm{m \, s^{-2}}$ should be given to 2 significant figures.

Test yourself

What to review

If your answer is incorrect:

1 A particle of mass 2 kg rests in equilibrium on a smooth plane which is inclined to the horizontal at an angle of 30°. The particle is kept in equilibrium by a force, of magnitude F, which acts in a vertical plane through a line of greatest slope of the plane. The force makes an angle of 40° with the plane, as shown in the diagram.

Review Edexcel Book M1 pages 93–95
Revise for M1 pages 33–38

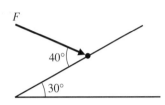

Find:

(a) the value of F,

(b) the magnitude of the normal reaction of the plane on the particle.

2 A particle of mass 1.5 kg is suspended by two light inextensible strings at angles of 40° and 60° with the horizontal. Find the tensions in the strings.

Review Edexcel Book M1 pages 97–100
Revise for M1 pages 33–38

3 A particle of mass 2.5 kg rests in limiting equilibrium on a rough inclined plane at an angle α to the horizontal, where $\tan \alpha = \frac{3}{4}$. The coefficient of friction between the particle and the plane is 0.8. A force of magnitude P acts on the particle along a line of greatest slope of the plane. Find P if the particle is on the point of slipping:

Review Edexcel Book M1 pages 103–108
Revise for M1 pages 33–38

(a) down the plane

(b) up the plane.

Worked examination question 1 [E]

Two horizontal forces **P** and **Q** act at a point O. The force **P** has magnitude 1 N and acts due north. The force **Q** has magnitude $3\sqrt{2}$ N and acts north-east. Calculate the magnitude and direction of the resultant of **P** and **Q**.

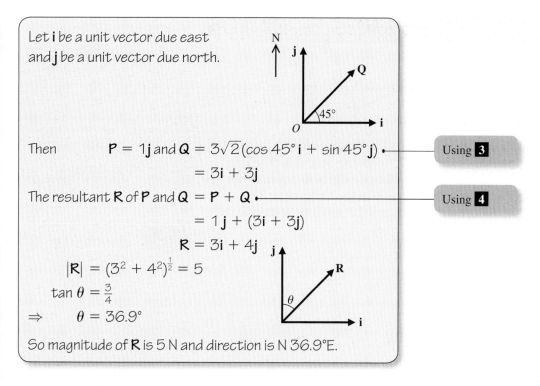

Let **i** be a unit vector due east
and **j** be a unit vector due north.

Then $\quad$ **P** $= 1\mathbf{j}$ and **Q** $= 3\sqrt{2}(\cos 45°\,\mathbf{i} + \sin 45°\,\mathbf{j})$ $\quad$ Using **3**

$$= 3\mathbf{i} + 3\mathbf{j}$$

The resultant **R** of **P** and **Q** $=$ **P** $+$ **Q** $\quad$ Using **4**

$$= 1\mathbf{j} + (3\mathbf{i} + 3\mathbf{j})$$

$$\mathbf{R} = 3\mathbf{i} + 4\mathbf{j}$$

$$|\mathbf{R}| = (3^2 + 4^2)^{\frac{1}{2}} = 5$$

$$\tan\theta = \frac{3}{4}$$

$$\Rightarrow \quad \theta = 36.9°$$

So magnitude of **R** is 5 N and direction is N 36.9°E.

Worked examination question 2 [E]

A horizontal force **R** is of magnitude 12 N and acts due east
from a point O. The horizontal forces **P** and **Q** act from O in
the directions 030° and due south respectively. Given that
P $+$ **Q** $=$ **R**, calculate the magnitudes of **P** and **Q**.

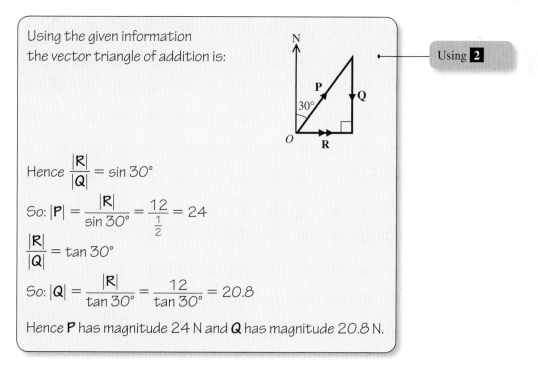

Using the given information
the vector triangle of addition is: $\quad$ Using **2**

Hence $\dfrac{|\mathbf{R}|}{|\mathbf{Q}|} = \sin 30°$

So: $|\mathbf{P}| = \dfrac{|\mathbf{R}|}{\sin 30°} = \dfrac{12}{\frac{1}{2}} = 24$

$\dfrac{|\mathbf{R}|}{|\mathbf{Q}|} = \tan 30°$

So: $|\mathbf{Q}| = \dfrac{|\mathbf{R}|}{\tan 30°} = \dfrac{12}{\tan 30°} = 20.8$

Hence **P** has magnitude 24 N and **Q** has magnitude 20.8 N.

4

Worked examination question 3 [E]

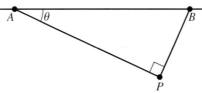

The figure shows a particle P of mass $0.5\,\text{kg}$ suspended by two light inextensible strings from two fixed points A and B. The line AB is horizontal and $\angle APB = 90°$. The particle is in equilibrium with the string AP inclined at an angle θ to the horizontal. The tensions in AP and BP are T newtons and $2T$ newtons respectively.

(a) Show that $\tan\theta = \frac{1}{2}$.

(b) Calculate, to 1 decimal place, the value of T.

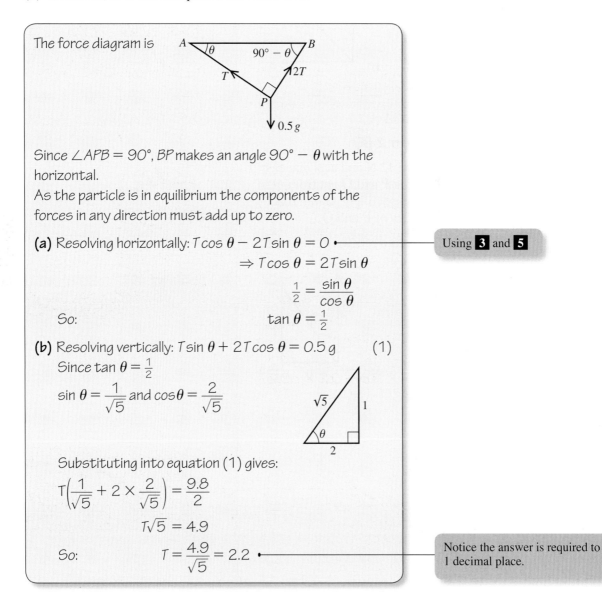

The force diagram is

Since $\angle APB = 90°$, BP makes an angle $90° - \theta$ with the horizontal.

As the particle is in equilibrium the components of the forces in any direction must add up to zero.

(a) Resolving horizontally: $T\cos\theta - 2T\sin\theta = 0$ ——— Using ③ and ⑤

$$\Rightarrow T\cos\theta = 2T\sin\theta$$

$$\frac{1}{2} = \frac{\sin\theta}{\cos\theta}$$

So: $$\tan\theta = \frac{1}{2}$$

(b) Resolving vertically: $T\sin\theta + 2T\cos\theta = 0.5\,g$ (1)

Since $\tan\theta = \frac{1}{2}$

$$\sin\theta = \frac{1}{\sqrt{5}} \text{ and } \cos\theta = \frac{2}{\sqrt{5}}$$

Substituting into equation (1) gives:

$$T\left(\frac{1}{\sqrt{5}} + 2 \times \frac{2}{\sqrt{5}}\right) = \frac{9.8}{2}$$

$$T\sqrt{5} = 4.9$$

So: $$T = \frac{4.9}{\sqrt{5}} = 2.2$$ ——— Notice the answer is required to 1 decimal place.

Worked examination question 4 [E]

A body of mass 6 kg is just prevented from sliding down a rough plane inclined at an angle α to the horizontal by a horizontal force of magnitude 39.2 N. The coefficient of friction between the plane and the body is 0.2.

(a) Find the angle α.

The horizontal force is now removed and the angle of inclination of the plane increased to the value θ where $\tan\theta = \frac{12}{5}$.

(b) Calculate the magnitude of the force **P** acting up the plane, along a line of greatest slope, which will just make the body start to move up the plane.

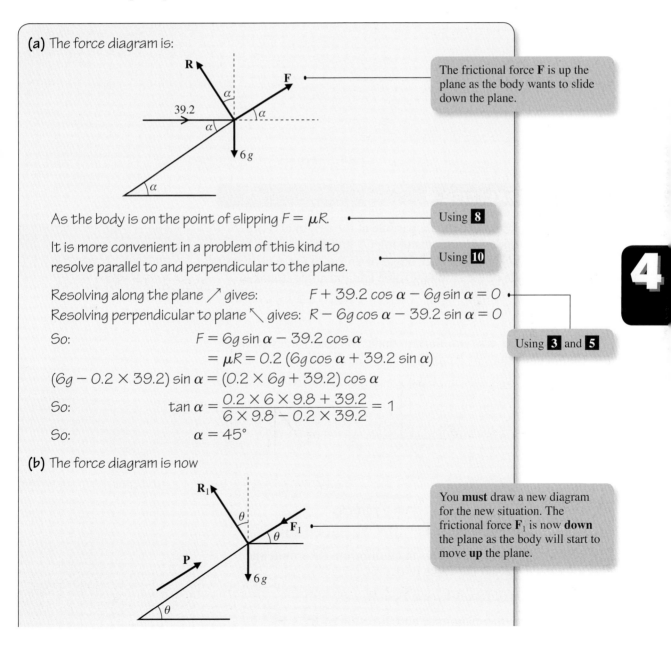

(a) The force diagram is:

The frictional force **F** is up the plane as the body wants to slide down the plane.

As the body is on the point of slipping $F = \mu R$. — Using **8**

It is more convenient in a problem of this kind to resolve parallel to and perpendicular to the plane. — Using **10**

Resolving along the plane ⟋ gives: $F + 39.2\cos\alpha - 6g\sin\alpha = 0$

Resolving perpendicular to plane ⟍ gives: $R - 6g\cos\alpha - 39.2\sin\alpha = 0$

So: $F = 6g\sin\alpha - 39.2\cos\alpha$

$= \mu R = 0.2\,(6g\cos\alpha + 39.2\sin\alpha)$ — Using **3** and **5**

$(6g - 0.2 \times 39.2)\sin\alpha = (0.2 \times 6g + 39.2)\cos\alpha$

So: $\tan\alpha = \dfrac{0.2 \times 6 \times 9.8 + 39.2}{6 \times 9.8 - 0.2 \times 39.2} = 1$

So: $\alpha = 45°$

(b) The force diagram is now

You **must** draw a new diagram for the new situation. The frictional force F_1 is now **down** the plane as the body will start to move **up** the plane.

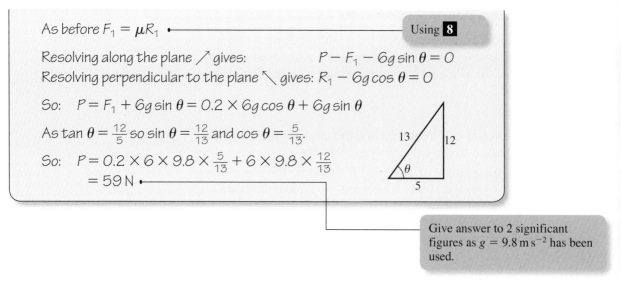

As before $F_1 = \mu R_1$ ●————————————————— Using **8**

Resolving along the plane ↗ gives: $\quad\quad P - F_1 - 6g \sin \theta = 0$

Resolving perpendicular to the plane ↘ gives: $R_1 - 6g \cos \theta = 0$

So: $\quad P = F_1 + 6g \sin \theta = 0.2 \times 6g \cos \theta + 6g \sin \theta$

As $\tan \theta = \frac{12}{5}$ so $\sin \theta = \frac{12}{13}$ and $\cos \theta = \frac{5}{13}$.

So: $\quad P = 0.2 \times 6 \times 9.8 \times \frac{5}{13} + 6 \times 9.8 \times \frac{12}{13}$

$\quad\quad = 59\,\text{N}$ ●

Give answer to 2 significant figures as $g = 9.8\,\text{m s}^{-2}$ has been used.

Revision exercise 4

1 Two forces $\mathbf{P}$ and $\mathbf{Q}$ act on a particle at the origin O. Force $\mathbf{P}$ has magnitude $4\,\text{N}$ and acts along Ox and force $\mathbf{Q}$ has magnitude $5\,\text{N}$ and acts at an angle of $75°$ with Ox. Calculate the magnitude of the resultant force and the angle it makes with Ox.

2 A particle at O is in equilibrium under the four forces shown. Obtain the values of P and Q.

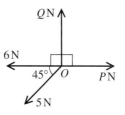

3 Three forces $\mathbf{F}_1$, $\mathbf{F}_2$ and $\mathbf{F}_3$ act on a particle.

$\mathbf{F}_1 = (-2\mathbf{i} + 5\mathbf{j})\,\text{N}$, $\mathbf{F}_2 = (\mathbf{i} - 2\mathbf{j})\,\text{N}$, $\mathbf{F}_3 = (p\mathbf{i} + q\mathbf{j})\,\text{N}$.

(a) Given that the particle is in equilibrium determine the value of p and the value of q.

The resultant of forces $\mathbf{F}_1$ and $\mathbf{F}_2$ is $\mathbf{R}$.

(b) Calculate the magnitude and direction of $\mathbf{R}$.

4 Three horizontal forces act on a particle at O as shown in the diagram.

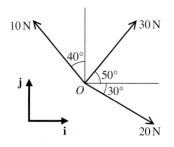

(a) Express the resultant force in terms of the unit vectors **i** and **j**.

(b) Find in terms of **i** and **j** the additional force that must also act if the particle is to be in equilibrium.

(c) Find the magnitude of this additional force.

5

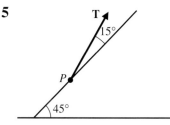

A particle P of mass 2 kg lies on a smooth plane inclined at 45° to the horizontal. The particle is held in equilibrium by a string which lies in a vertical plane through a line of greatest slope and makes an angle of 15° with the plane as shown in the figure. The tension in the string is **T** and the force exerted by the plane on the particle is **R**. Find **T** and **R**.

6 A particle P of mass 2 kg is placed on a rough horizontal table. The coefficient of friction between P and the table is μ. A force of magnitude 6 N acting upwards at an angle of 60° to the horizontal is applied to P. Given that equilibrium is on the point of being broken by the particle sliding on the table find, to 1 decimal place, the value of μ.

4

Moments

What you should know

1 The **moment** of a force of magnitude F about a point P is given by:

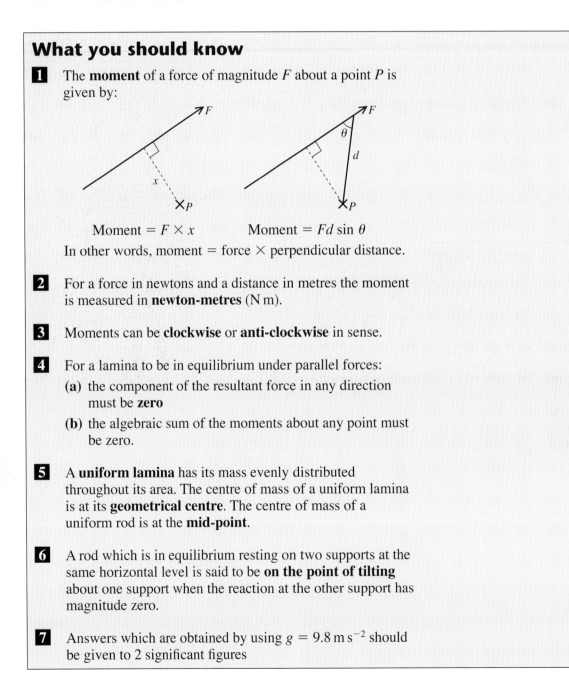

Moment $= F \times x$ Moment $= Fd \sin \theta$

In other words, moment $=$ force $\times$ perpendicular distance.

2 For a force in newtons and a distance in metres the moment is measured in **newton-metres** (N m).

3 Moments can be **clockwise** or **anti-clockwise** in sense.

4 For a lamina to be in equilibrium under parallel forces:
  (a) the component of the resultant force in any direction must be **zero**
  (b) the algebraic sum of the moments about any point must be zero.

5 A **uniform lamina** has its mass evenly distributed throughout its area. The centre of mass of a uniform lamina is at its **geometrical centre**. The centre of mass of a uniform rod is at the **mid-point**.

6 A rod which is in equilibrium resting on two supports at the same horizontal level is said to be **on the point of tilting** about one support when the reaction at the other support has magnitude zero.

7 Answers which are obtained by using $g = 9.8 \, \text{m s}^{-2}$ should be given to 2 significant figures

If your answer is incorrect:

1 A uniform rod of length 1.2 m and mass 3 kg is placed on a horizontal table with part of its length protruding over the edge of the table. When a parcel of mass 0.6 kg is suspended from the free end of the rod, the rod is on the point of tilting. Calculate the length of the section of the rod which rests on the table.

Review Edexcel Book M1 page 124
Revise for M1 pages 40–46

2 A uniform horizontal beam ABC, of mass 10 kg and length 2 m, rests in equilibrium on two supports, one at A and the other at B, where $AB = 1.7$ m. A load of mass 12 kg is hung from C. Find the magnitude, in N, of the reaction at each support.   [E]

Review Edexcel Book M1 pages 123–124
Revise for M1 pages 40–46

3 A uniform straight plank AB, of mass 12 kg and length 2 m, rests horizontally on two supports, one at C and the other at D, where $AC = CD = 0.6$ m. A particle P of mass X kg is hung from B and the plank is on the point of tilting.

(a) Find the value of X.

The particle P is removed from B and hung from A.

(b) Find, in N, the magnitude of the force exerted on the plank at each support.   [E]

Review Edexcel Book M1 pages 123–124
Revise for M1 pages 40–46

4 A non-uniform plank PQ of mass 5 kg and length 4 m is in equilibrium in a horizontal position resting on supports at points S and T of the plank where $PS = QT = 1$ m. When a block of mass $3M$ kg is placed at Q the plank is on the point of tilting. If instead a block of mass $2M$ kg is placed at P the plank is also on the point of tilting. Find:

(a) the value of M

(b) the distance of the centre of mass from P.

Review Edexcel Book M1 pages 126–128
Revise for M1 pages 40–46

Example 1

A non-uniform rod AB of length 3 m and mass 5 kg is suspended in equilibrium in a horizontal position by ropes attached to points P and Q of the rod. $AP = 1$ m and $AQ = 2.5$ m. The tensions in the ropes are equal. Find the distance of the centre of mass of the rod from A.

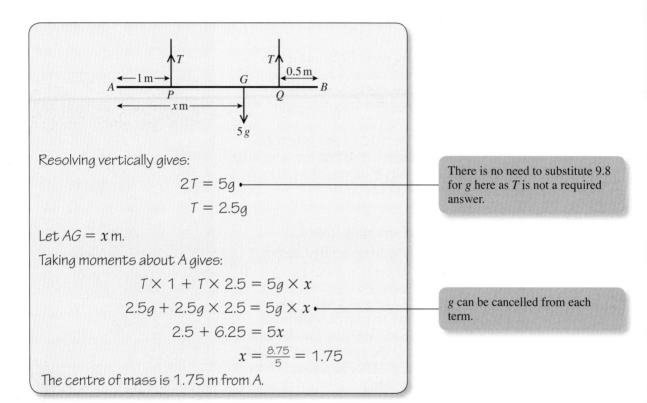

Resolving vertically gives:

$$2T = 5g$$

$$T = 2.5g$$

There is no need to substitute 9.8 for g here as T is not a required answer.

Let $AG = x$ m.

Taking moments about A gives:

$$T \times 1 + T \times 2.5 = 5g \times x$$

$$2.5g + 2.5g \times 2.5 = 5g \times x$$

g can be cancelled from each term.

$$2.5 + 6.25 = 5x$$

$$x = \frac{8.75}{5} = 1.75$$

The centre of mass is 1.75 m from A.

Example 2

A uniform plank AB of length 6 m and mass 10 kg rests in equilibrium in a horizontal position on supports at points P and Q of the plank. $AP = 2$ m and $AQ = 4.8$ m. When a box of mass M kg is placed on the plank at D, where $AD = 0.5$ m, the plank is on the point of tilting.

(a) Suggest suitable models for the plank and the box.

(b) Calculate, to 3 significant figures, the value of M.

(a) The plank can be modelled as a uniform rod.
    The box can be modelled as a particle.

(b)

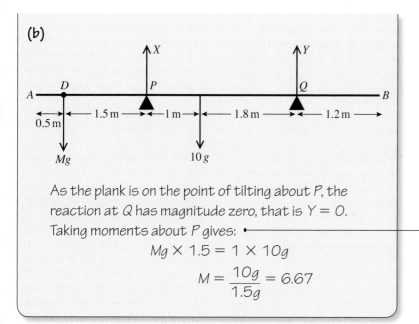

As the plank is on the point of tilting about P, the reaction at Q has magnitude zero, that is $Y = 0$.

Taking moments about P gives:

$$Mg \times 1.5 = 1 \times 10g$$

$$M = \frac{10g}{1.5g} = 6.67$$

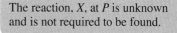

The reaction, X, at P is unknown and is not required to be found.

Worked examination question 1 [E]

A uniform plank AB of length 4 m and weight 70 N, rests horizontally on supports at P and Q, where $PQ = 2$ m and P lies between A and Q. A block of weight 30 N hangs from the plank at A. Given that the forces on the supports at P and Q are equal in magnitude, find the distance AP.

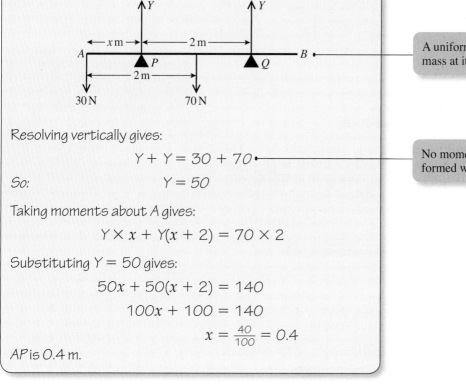

A uniform plank has its centre of mass at its mid-point.

Resolving vertically gives:

$$Y + Y = 30 + 70$$

So:

$$Y = 50$$

No moments equation can be formed without Y, so find Y first.

Taking moments about A gives:

$$Y \times x + Y(x + 2) = 70 \times 2$$

Substituting $Y = 50$ gives:

$$50x + 50(x + 2) = 140$$

$$100x + 100 = 140$$

$$x = \frac{40}{100} = 0.4$$

AP is 0.4 m.

Worked examination question 2 [E]

A straight uniform rigid rod AB is of length 8 m and mass 10 kg. The rod is supported at the point X, where AX = 5m, and, when downward vertical forces of magnitudes P and 4P newtons are applied at A and B respectively, the rod rests in equilibrium with AB horizontal. Calculate:

(a) the value of P

(b) the force, in N, exerted on the support at X.

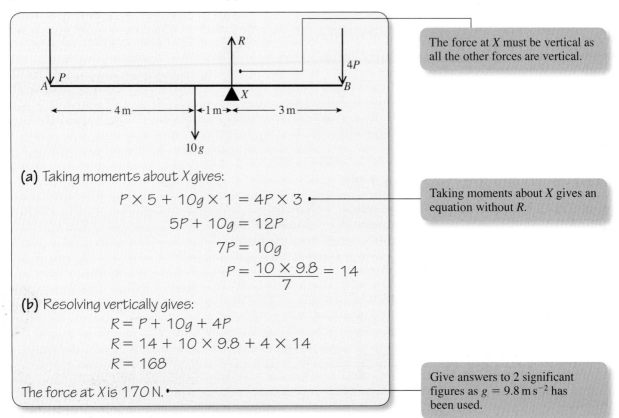

The force at X must be vertical as all the other forces are vertical.

(a) Taking moments about X gives:

$$P \times 5 + 10g \times 1 = 4P \times 3$$

$$5P + 10g = 12P$$

$$7P = 10g$$

$$P = \frac{10 \times 9.8}{7} = 14$$

Taking moments about X gives an equation without R.

(b) Resolving vertically gives:

$$R = P + 10g + 4P$$

$$R = 14 + 10 \times 9.8 + 4 \times 14$$

$$R = 168$$

The force at X is 170 N.

Give answers to 2 significant figures as g = 9.8 m s⁻² has been used.

Worked examination question 3 [E]

A smooth uniform rod AB is 3 m long and has weight 30 N. The rod is supported in a horizontal position by a trestle at Q, 70 cm from B, and by a rope passing under the rod at P, 80 cm from A. Both parts of the rope supporting the rod are vertical.

(a) Calculate the values of the vertical reaction at Q and the tension in the rope, assuming that this tension is constant throughout that part of the rope in contact with the rod.

A body of weight 50 N is now suspended from the rod AB at a point X.

(b) Calculate the least value of the distance AX if the rod is not to lose contact with the trestle at Q.

(c) If X is between P and Q, calculate the length of PX when the tension in the rope is equal to the reaction at Q.

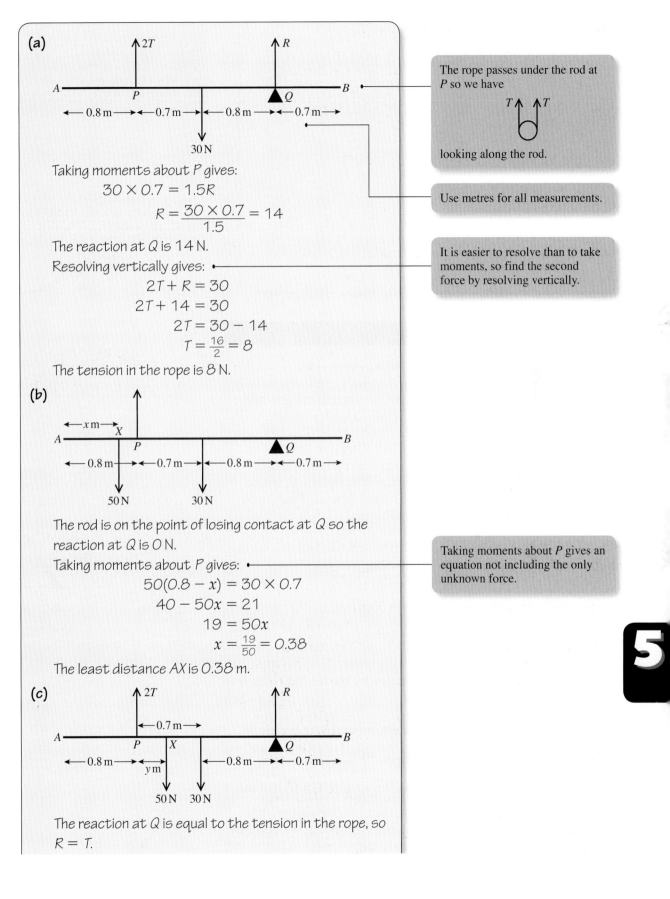

(a)

2T   R

A ————————————————— B
    P   ▲Q
◄—0.8 m—►◄—0.7 m—►◄—0.8 m—►◄—0.7 m—►

30 N

The rope passes under the rod at P so we have

T↑ ↑T

looking along the rod.

Taking moments about P gives:
$$30 \times 0.7 = 1.5R$$
$$R = \frac{30 \times 0.7}{1.5} = 14$$

The reaction at Q is 14 N.

Use metres for all measurements.

Resolving vertically gives:

It is easier to resolve than to take moments, so find the second force by resolving vertically.

$$2T + R = 30$$
$$2T + 14 = 30$$
$$2T = 30 - 14$$
$$T = \frac{16}{2} = 8$$

The tension in the rope is 8 N.

(b)

◄—x m—►
        X
A ————————————————— B
    P   ▲Q
◄—0.8 m—►◄—0.7 m—►◄—0.8 m—►◄—0.7 m—►

50 N   30 N

The rod is on the point of losing contact at Q so the reaction at Q is 0 N.

Taking moments about P gives:

Taking moments about P gives an equation not including the only unknown force.

$$50(0.8 - x) = 30 \times 0.7$$
$$40 - 50x = 21$$
$$19 = 50x$$
$$x = \frac{19}{50} = 0.38$$

The least distance AX is 0.38 m.

(c)

2T   R
◄—0.7 m—►
A ————————————————— B
    P X   ▲Q
◄—0.8 m—►◄ ►◄—0.8 m—►◄—0.7 m—►
       y m

50 N 30 N

The reaction at Q is equal to the tension in the rope, so R = T.

Resolving vertically gives:
$$2T + R = 50 + 30$$
$$2T + T = 80$$
$$T = \frac{80}{3}$$

So:
$$R = \frac{80}{3}$$

Taking moments about P gives:
$$50y + 30 \times 0.7 = R \times 1.5$$
$$50y + 21 = \frac{80}{3} \times 1.5$$
$$50y = \frac{80}{3} \times 1.5 - 21$$
$$y = \frac{19}{50} = 0.38$$

The length of PX is 0.38 m.

Any moments equation will include one or both of the unknown forces, so resolve first to find the tension and hence the reaction at Q.

Revision exercise 5

1 A straight uniform beam ABCD, where AB = 1 m, BC = 2.5 m, CD = 0.5 m, rests horizontally on supports at B and C. The beam is of mass 10 kg. Calculate the magnitude, in N, of the force exerted by the beam on each support.   [E]

2 A uniform straight beam AB, of mass 50 kg and length 8 m, is suspended horizontally by two vertical light inextensible ropes, one attached at A and the other at the point C, where AC = 7 m. A load of mass 20 kg is hung from B. Find, in N, the magnitude of the tension in each rope.   [E]

3 A uniform straight beam ABCD, of length 6 m and mass 20 kg, is attached to a fixed hinge at C, where
$$AB = BC = CD.$$
The beam, which can turn freely in a vertical plane, remains at rest with ABCD horizontal when bodies of masses 10 kg, x kg and 55 kg are hung from A, B and D respectively.

(a) Calculate the value of x.

(b) Determine, in N, the force exerted by the loaded beam on the hinge.   [E]

4 The centre of mass of a non-uniform beam AB, of weight W and length 4 m, is at the point G. The beam rests horizontally on two supports X and Y, where AX = 1 m and YB = 2 m.

A weight $\frac{W}{4}$ is suspended from the beam at B. Given that the magnitude of the force exerted on the beam by support Y is three times the magnitude of the force exerted on the beam by support X, calculate the distance AG.   [E]

5 A plank AB of length 6.5 m and weight 800 N is supported horizontally by two vertical forces acting at points X and Y on the plank. $AX = 0.8$ m, $YB = 2.1$ m and the centre of mass of the plank is at G, where $AG = 3.5$ m. A boy of weight 450 N stands at A and then walks slowly towards B. Calculate:

(a) the forces at Y and X when the boy is standing at A

(b) the forces at Y and X when the boy is standing at the point Y

(c) the distance the boy has walked from Y when the plank starts to tip. [E]

6 A uniform rod, AB, 1.44 m long and weighing 50 N, carries at B a body of weight 40 N. It is maintained in a horizontal position by an upward vertical force at a variable point X. Calculate the distance XB:

(a) when the force at A is 12 N (acting upwards)

(b) when the force at A is zero.

Determine also the distance XB when the force at A is equal to its greatest value. What is then the magnitude of the force at X? [E]

5

6

What you should know

1 **Scalar quantities** are completely specified by their magnitude.

2 **Vector quantities** require both their magnitude and direction to be specified.

3 Vectors are added by using the triangle law of addition

$$\overrightarrow{OB} = \overrightarrow{OA} + \overrightarrow{AB}$$

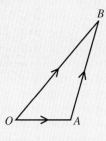

4 Using the **i**, **j** notation

$$\mathbf{R} = X\mathbf{i} + Y\mathbf{j}$$

where $X = R\cos\theta$, $Y = R\sin\theta$
and R = magnitude of $\mathbf{R} = |\mathbf{R}| = (X^2 + Y^2)^{\frac{1}{2}}$
X and Y are the **components** or **resolutes** of **R**.

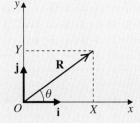

5 If $\mathbf{a} = a_1\mathbf{i} + a_2\mathbf{j}$ and $\mathbf{b} = b_1\mathbf{i} + b_2\mathbf{j}$ then
$$\mathbf{a} + \mathbf{b} = (a_1 + b_1)\mathbf{i} + (a_2 + b_2)\mathbf{j}$$
$$\mathbf{a} - \mathbf{b} = (a_1 - b_1)\mathbf{i} + (a_2 - b_2)\mathbf{j}$$

6 If particle A has position vector $\mathbf{r}_A$ and particle B has position vector $\mathbf{r}_B$ then the position vector of B relative to A is $\mathbf{r}_B - \mathbf{r}_A$.

7 If particle A has velocity $\mathbf{v}_A$ and particle B has velocity $\mathbf{v}_B$ then the velocity of B relative to A is $\mathbf{v}_B - \mathbf{v}_A$.

8 If the **velocity** of a particle is **constant** then

change of displacement = velocity × time

9 If the **acceleration** of a particle is **constant** then

change of velocity = acceleration × time.

1 Two forces $\mathbf{F}_1$ and $\mathbf{F}_2$ act on a particle. The force $\mathbf{F}_1$ is parallel to the vector $(\mathbf{i} + \mathbf{j})$ and the force $\mathbf{F}_2$ is parallel to the vector $(\mathbf{i} + 2\mathbf{j})$. Given that the resultant $\mathbf{R}$ of these two forces is of magnitude 15 N and parallel to the vector $(3\mathbf{i} + 4\mathbf{j})$, find the magnitude of $\mathbf{F}_1$ and the magnitude of $\mathbf{F}_2$.

Review Edexcel Book M1 pages 140–149 Revise for M1 pages 48–51

2 An aeroplane flies from airport A to airport B which is 100 km away on a bearing of 060°. The aeroplane then flies to airport C which is 120 km from B on a bearing of 195°.
   (a) Find the distance from A to C.
   (b) Find the course the aeroplane must fly to return to A from C.

Review Edexcel Book M1 pages 134–135 Revise for M1 pages 48–51

3 At 12 noon the position vectors $\mathbf{r}$ and the velocity vectors $\mathbf{v}$ of two ships A and B are:
$\mathbf{r}_A = (11\mathbf{i} + 13\mathbf{j})\,\text{km}$ $\mathbf{v}_A = (6\mathbf{i} + 12\mathbf{j})\,\text{km h}^{-1}$
$\mathbf{r}_B = (18\mathbf{i} + 17\mathbf{j})\,\text{km}$ $\mathbf{v}_B = (-8\mathbf{i} + 4\mathbf{j})\,\text{km h}^{-1}.$
Show that if the ships do not alter their velocities a collision will occur. Find the time when the collision occurs and the position vector of the location of the collision.

Review Edexcel Book M1 pages 143–148 Revise for M1 pages 48–51

Example 1

A girl walks 5 km due east from O to A then 3 km in a north-east direction from A to B. Find the distance of B from O and describe the displacement $\overrightarrow{OB}$.

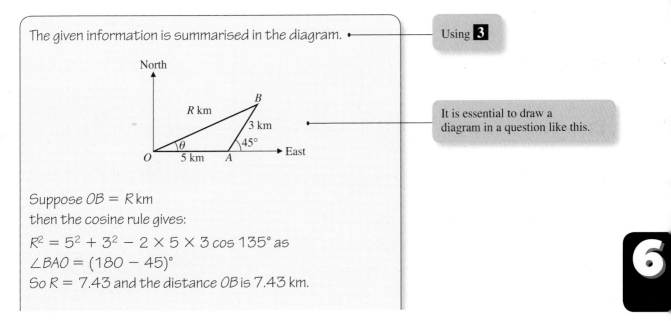

The given information is summarised in the diagram.

Using **3**

It is essential to draw a diagram in a question like this.

Suppose $OB = R\,\text{km}$
then the cosine rule gives:
$R^2 = 5^2 + 3^2 - 2 \times 5 \times 3 \cos 135°$ as
$\angle BAO = (180 - 45)°$
So $R = 7.43$ and the distance OB is 7.43 km.

6

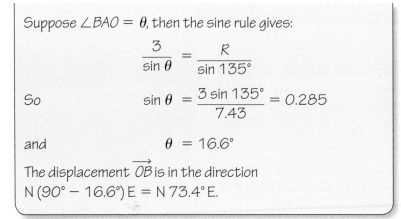

Suppose $\angle BAO = \theta$, then the sine rule gives:

$$\frac{3}{\sin \theta} = \frac{R}{\sin 135°}$$

So

$$\sin \theta = \frac{3 \sin 135°}{7.43} = 0.285$$

and

$$\theta = 16.6°$$

The displacement $\overrightarrow{OB}$ is in the direction
$N (90° - 16.6°) E = N 73.4° E$.

Example 2

At a given time particle A has position vector $(5\mathbf{i} - 4\mathbf{j})$ m relative to a fixed origin O and particle B has position vector $(6\mathbf{i} + 4\mathbf{j})$ m relative to O. Find the position vector of B relative to A.

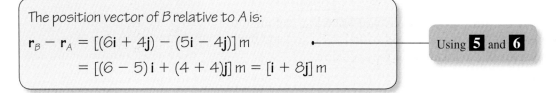

The position vector of B relative to A is:

$\mathbf{r}_B - \mathbf{r}_A = [(6\mathbf{i} + 4\mathbf{j}) - (5\mathbf{i} - 4\mathbf{j})]$ m

$\qquad = [(6 - 5)\mathbf{i} + (4 + 4)\mathbf{j}]$ m $= [\mathbf{i} + 8\mathbf{j}]$ m

Using **5** and **6**

Example 3

At a given instant a small parcel is sliding down a slope inclined at $30°$ to the horizontal, with a speed of 10 m s^{-1}. Find $\mathbf{v}$, the velocity of the parcel, in the form $\alpha\mathbf{i} + \beta\mathbf{j}$ where $\mathbf{i}$ and $\mathbf{j}$ are unit vectors horizontally and vertically upwards respectively.

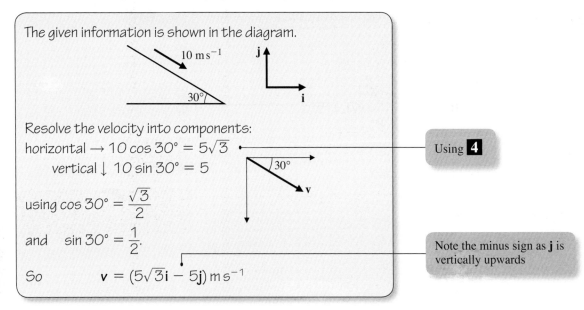

The given information is shown in the diagram.

Resolve the velocity into components:
horizontal $\rightarrow 10 \cos 30° = 5\sqrt{3}$
$\quad$ vertical $\downarrow 10 \sin 30° = 5$

using $\cos 30° = \frac{\sqrt{3}}{2}$

and $\quad \sin 30° = \frac{1}{2}$.

So $\qquad \mathbf{v} = (5\sqrt{3}\mathbf{i} - 5\mathbf{j}) \text{ m s}^{-1}$

Using **4**

Note the minus sign as $\mathbf{j}$ is vertically upwards

Example 4

At time $t = 0$ a particle P has velocity $(2\mathbf{i} + 3\mathbf{j})\,\mathrm{m\,s^{-1}}$.
Two seconds later its velocity is $(14\mathbf{i} + 9\mathbf{j})\,\mathrm{m\,s^{-1}}$. Find the acceleration of P given that it is constant. Obtain the magnitude and direction of this acceleration.

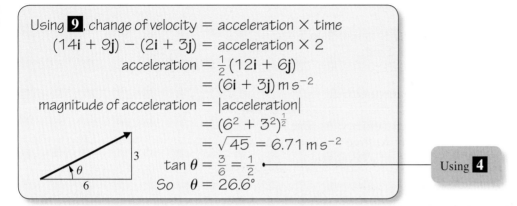

Using **9**, change of velocity = acceleration × time

$$(14\mathbf{i} + 9\mathbf{j}) - (2\mathbf{i} + 3\mathbf{j}) = \text{acceleration} \times 2$$
$$\text{acceleration} = \tfrac{1}{2}(12\mathbf{i} + 6\mathbf{j})$$
$$= (6\mathbf{i} + 3\mathbf{j})\,\mathrm{m\,s^{-2}}$$
$$\text{magnitude of acceleration} = |\text{acceleration}|$$
$$= (6^2 + 3^2)^{\frac{1}{2}}$$
$$= \sqrt{45} = 6.71\,\mathrm{m\,s^{-2}}$$
$$\tan\theta = \tfrac{3}{6} = \tfrac{1}{2}$$
$$\text{So} \quad \theta = 26.6°$$

Using **4**

Example 5

(In this question the unit vectors $\mathbf{i}$ and $\mathbf{j}$ are due east and due north respectively.)

At 12 noon the position vectors $\mathbf{r}$ and the velocity vectors $\mathbf{v}$ of two boats A and B are

$\mathbf{r}_A = (\mathbf{i} + 3\mathbf{j})\,\mathrm{km}$ $\mathbf{v}_A = (2\mathbf{i} + \mathbf{j})\,\mathrm{km\,h^{-1}}$
$\mathbf{r}_B = (\mathbf{i} + \mathbf{j})\,\mathrm{km}$ $\mathbf{v}_B = (7\mathbf{i} + 5\mathbf{j})\,\mathrm{km\,h^{-1}}$

(a) Show that at time t hours after noon the position vector of B relative to A is given by $[5t\mathbf{i} + (4t - 2)\mathbf{j}]\,\mathrm{km}$.

(b) Hence find the time at which B is due east of A. How far apart are they then?

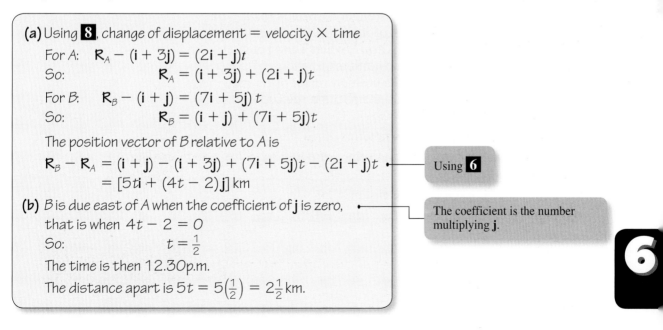

(a) Using **8**, change of displacement = velocity × time

For A: $R_A - (\mathbf{i} + 3\mathbf{j}) = (2\mathbf{i} + \mathbf{j})t$
So: $R_A = (\mathbf{i} + 3\mathbf{j}) + (2\mathbf{i} + \mathbf{j})t$

For B: $R_B - (\mathbf{i} + \mathbf{j}) = (7\mathbf{i} + 5\mathbf{j})t$
So: $R_B = (\mathbf{i} + \mathbf{j}) + (7\mathbf{i} + 5\mathbf{j})t$

The position vector of B relative to A is

$$R_B - R_A = (\mathbf{i} + \mathbf{j}) - (\mathbf{i} + 3\mathbf{j}) + (7\mathbf{i} + 5\mathbf{j})t - (2\mathbf{i} + \mathbf{j})t$$
$$= [5t\mathbf{i} + (4t - 2)\mathbf{j}]\,\mathrm{km}$$

Using **6**

(b) B is due east of A when the coefficient of $\mathbf{j}$ is zero, that is when $4t - 2 = 0$
So: $t = \tfrac{1}{2}$

The coefficient is the number multiplying $\mathbf{j}$.

The time is then 12.30 p.m.

The distance apart is $5t = 5\left(\tfrac{1}{2}\right) = 2\tfrac{1}{2}\,\mathrm{km}$.

6

Revision exercise 6

1   Bhavana walks 5 km due east from O to A and then 6 km due north to B. Find:

   (a) the distance of B from O

   (b) the direction of the displacement OB.

2   Bill walks 4 km due west from O to A and then 5 km in a north-east direction from A to B. Find:

   (a) the distance of B from O

   (b) the direction of the displacement OB.

3   A particle P moves with constant velocity. At time $t = 0$ its position vector is $(3\mathbf{i} - 4\mathbf{j})$ m. At time $t = 3$ seconds its position vector is $(6\mathbf{i} + 5\mathbf{j})$ m. Find its constant velocity.

4   A particle P has velocity $(4\mathbf{i} + 7\mathbf{j})$ m s^{-1} initially and velocity $(9\mathbf{i} - 2\mathbf{j})$ m s^{-1} 5 seconds later. Find its acceleration, assumed constant, and the magnitude of that acceleration.

5   A helicopter A leaves a heliport O and flies with velocity $(9\mathbf{i} + 8\mathbf{j})$ m s^{-1}. At the same time a helicopter B takes off from another heliport whose position vector, relative to O, is $(25\mathbf{i} + 12\mathbf{j})$ m. The velocity of B is $(-5\mathbf{i} + 6\mathbf{j})$ m s^{-1}. Find after t seconds:

   (a) the position vector of A relative to O

   (b) the position vector of B relative to O

   (c) the position vector of B relative to A.

   (d) Hence show that the helicopters do not collide.

6   Two cars A and B are moving on straight horizontal roads with constant velocities. The velocity of A is 15 m s^{-1} due east, and the velocity of B is $20(\mathbf{i} - \mathbf{j})$ m s^{-1}, where $\mathbf{i}$ and $\mathbf{j}$ are unit vectors directed due east and due north respectively. When $t = 0$, A is at the origin O and the position vector of B relative to O is $200\mathbf{j}$ m. At time t seconds, the position vectors of A and B are $\mathbf{r}$ and $\mathbf{s}$.

   (a) Find expressions for $\mathbf{r}$ and $\mathbf{s}$ in terms of t.

   (b) Hence write down the position vector of B relative to A in terms of t.

   (c) Find the time when B is due east of A.

   (d) Find the time when the cars are again 200 m apart.

Examination style paper

Attempt all questions   **Time allowed 90 minutes**

Whenever a numerical value of g is required, take $g = 9.8\,\text{ms}^{-2}$

1. Two forces $(6\mathbf{i} + 7\mathbf{j})\,\text{N}$ and $(3\mathbf{i} + 5\mathbf{j})\,\text{N}$ act on a particle P.
   (a) Find the resultant of these two forces.   **(2)**

   Given that P has mass 1.5 kg,
   (b) find the magnitude of the acceleration of P.   **(4)**

2. Two small smooth spheres P and Q have mass m_1 and m_2 respectively. They are moving towards each other along the same horizontal line with speed u. After the collision both spheres have reversed their original direction of motion. Given that P travels with speed $\frac{1}{2}u$ show that $3m_1 > 2m_2$. **(6)**

3. A pole-vaulter uses a uniform pole AB of length 4 m and mass 4.5 kg. She holds the pole horizontally by placing one hand at one end A of the pole and the other hand at the point C where $AC = 70$ cm. Find the vertical forces exerted by her hands on the pole at A and C.   **(7)**

4. (In this question the horizontal unit vectors $\mathbf{i}$ and $\mathbf{j}$ are directed due east and due north respectively.)

   At 1 p.m. car A is at $10\mathbf{j}$ km and car B is at $30\mathbf{i}$ km. Car A moves with speed $64\,\text{km h}^{-1}$ due east and car B moves with velocity $(40\mathbf{i} + 8\mathbf{j})\,\text{km h}^{-1}$.

   (a) Show that the cars will collide.   **(6)**
   (b) Find the time at which the collision will take place and the position vector of the point of collision.   **(3)**

5. The figure shows a metal ball of mass 5 kg suspended by two chains from points A and B on the same horizontal level as each other and $1\frac{3}{4}$ m apart. The ball rests in equilibrium with P 1 m below AB and $\angle PAB = 45°$.

   By modelling P as a particle and the chains by light inextensible strings, find the tensions in PA and PB.   **(10)**

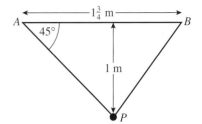

6 A train starts from rest at Ambury and moves with constant acceleration for 80 s until it reaches a speed of 20 m s^{-1}. It travels at this constant speed for T seconds. The train then decelerates for a distance of 400 m before coming to rest at Bexbury. The distance from Ambury to Bexbury is 8 km.

    (a) Sketch a speed–time graph for the journey. **(2)**

    (b) Calculate the deceleration of the train. **(3)**

    (c) Calculate the value of T. **(3)**

    (d) Calculate the total time for the journey. **(2)**

7 Two particles A and B are connected by a light inextensible string which passes over a smooth fixed pulley. The hanging parts of the string are vertical. Particle A has mass 6 kg and descends 9.8 m in the first two seconds after the system is released from rest.

    (a) Show that the acceleration of A is 4.9 m s^{-2}. **(3)**

    (b) Find the mass of particle B. **(6)**

    (c) Find the tension in the string. **(2)**

8 A particle P is at rest at a point A on a rough plane inclined at an angle $\arctan \frac{3}{4}$ to the horizontal. The coefficient of friction between the particle and the plane is $\frac{1}{4}$. The particle is projected up a line of greatest slope of the plane from the point A with a speed of 20 m s^{-1} and comes to instantaneous rest at the point B.

    (a) Show that, while P is moving up the plane, its acceleration is of magnitude $\frac{4}{5}g$ and is directed down the plane. **(7)**

    (b) Find the distance AB in metres to 3 significant figures. **(3)**

    (c) Find the time taken for P to move from A to B, giving your answer in seconds to 3 significant figures. **(3)**

    (d) Find the speed in m s^{-1} of the particle when it returns to A. **(4)**

Answers

Revision exercise 1

1 Golf ball, snooker ball, cricket ball, the Earth in its motion relative to the Sun.

2 Particle on a smooth plane pulled by an inextensible string and no air resistance. Could take into account: size of box, where rope is attached, the roughness of the playground, air resistance, the fact that the rope will extend during the motion.

Revision exercise 2

1 (a) $1.5\,\text{m s}^{-2}$ **(b)** $26\tfrac{2}{3}\,\text{s}$

2 (a) $11\,\text{m}$ **(b)** $15\,\text{m s}^{-1}$
(c) $2.5\,\text{s}$

3 (a) $1306\,\text{m}$
(b) $47.5\,\text{m s}^{-1}$

4 (a) $39\,\text{m s}^{-1}$ **(b)** $5.2\,\text{s}$

5 (a) $25\,\text{s}$ **(b)** $206\,\text{m}$

6 (a) $28\,\text{m s}^{-1}$ **(b)** $2.9\,\text{s}$

7 $16\,\text{m}$

8 (a) $25.6\,\text{s}$ **(b)** $513\,\text{m}$

9 (a)

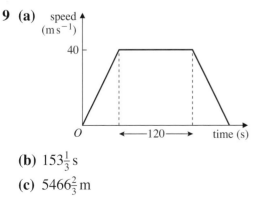

(b) $153\tfrac{1}{3}\,\text{s}$
(c) $5466\tfrac{2}{3}\,\text{m}$

10 (a) $v = 5f$
(b)

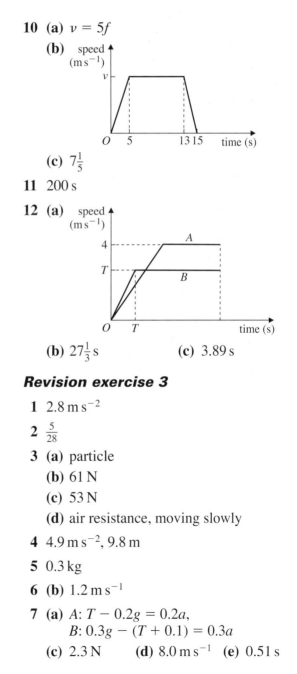

(c) $7\tfrac{1}{5}$

11 $200\,\text{s}$

12 (a)

(b) $27\tfrac{1}{3}\,\text{s}$ **(c)** $3.89\,\text{s}$

Revision exercise 3

1 $2.8\,\text{m s}^{-2}$

2 $\tfrac{5}{28}$

3 (a) particle
(b) $61\,\text{N}$
(c) $53\,\text{N}$
(d) air resistance, moving slowly

4 $4.9\,\text{m s}^{-2}, 9.8\,\text{m}$

5 $0.3\,\text{kg}$

6 (b) $1.2\,\text{m s}^{-1}$

7 (a) $A: T - 0.2g = 0.2a,$
$\quad\ B: 0.3g - (T + 0.1) = 0.3a$
(c) $2.3\,\text{N}$ **(d)** $8.0\,\text{m s}^{-1}$ **(e)** $0.51\,\text{s}$

8 (a) $1.4\,\mathrm{m\,s^{-2}}$ (b) $1.4\,\mathrm{m\,s^{-2}}$ (c) $3.4\,\mathrm{N}$

9 $M = 4000,\ \lambda = \frac{1}{4}$

10 (a) $2120\,\mathrm{N\,s}$ (b) $10\,600\,\mathrm{N}$

11 $6\,\mathrm{m\,s^{-1}}$

12 $5.6\,\mathrm{N\,s}$

13 $0.005\,\mathrm{kg}$

14 (a) u (b) $\frac{u}{2}$ (c) $\frac{3mu}{2}$

15 (a) $4u$ (b) $6mu$

16 (a) $\frac{g}{3}$ (b) $1.81\,\mathrm{m\,s^{-1}}$
(c) $0.602\,\mathrm{m\,s^{-1}}$ (d) $0.121\,\mathrm{N\,s}$

17 (a) $\frac{g}{11}\,\mathrm{m\,s^{-2}}$ (c) $1.5\,\mathrm{s}$

18 $1.75\,\mathrm{m\,s^{-1}}$

Revision exercise 4

1 $7.17\,\mathrm{N},\ 42.4°$

2 $P = 9.54,\ Q = 3.54$

3 (a) $p = 1,\ q = -3$
(b) $3.16\,\mathrm{N},\ 108°$ with $\mathbf{i}$

4 (a) $30.2\mathbf{i} + 20.6\mathbf{j}$ (b) $-30.2\mathbf{i} - 20.6\mathbf{j}$
(c) $36.6\,\mathrm{N}$

5 $\mathbf{T} = 14\,\mathrm{N},\ \mathbf{R} = 10\,\mathrm{N}$

6 $\mu = 0.2$

Revision exercise 5

1 $59\,\mathrm{N},\ 39\,\mathrm{N}$

2 $500\,\mathrm{N},\ 180\,\mathrm{N}$

3 (a) 25 (b) $1100\,\mathrm{N}$

4 $1\frac{3}{16}\,\mathrm{m}$ or $1.19\,\mathrm{m}$

5 (a) $750\,\mathrm{N},\ 500\,\mathrm{N}$
(b) $200\,\mathrm{N},\ 1050\,\mathrm{N}$
(c) $6\,\mathrm{m}$

6 (a) $0.24\,\mathrm{m}$ (b) $0.4\,\mathrm{m};\ 0\,\mathrm{m},\ 65\,\mathrm{N}$

Revision exercise 6

1 (a) $7.81\,\mathrm{km}$
(b) N 39.8° E or 039.8°

2 (a) $3.57\,\mathrm{km}$
(b) N 7.5° W (bearing of 352.5)

3 $(\mathbf{i} + 3\mathbf{j})\,\mathrm{m\,s^{-1}}$

4 $(\mathbf{i} - 1.8\mathbf{j})\,\mathrm{m\,s^{-2}},\ 2.06\,\mathrm{m\,s^{-2}}$

5 (a) $(9\mathbf{i} + 8\mathbf{j})t\,\mathrm{m}$
(b) $[(25\mathbf{i} + 12\mathbf{j}) + (\{-\}5\mathbf{i} + 6\mathbf{j})t]\,\mathrm{m}$
(c) $[(25 - 14t)\mathbf{i} + (12 - 2t)\mathbf{j}]\,\mathrm{m}$

6 (a) $\mathbf{r} = 15t\mathbf{i},\ \mathbf{s} = 20t\mathbf{i} + (200 - 20t)\mathbf{j}$
(b) $5t\mathbf{i} + (200 - 20t)\mathbf{j}$
(c) $t = 10$
(d) $18.8\,\mathrm{s}$

Examination style paper

1 (a) $(9\mathbf{i} + 12\mathbf{j})\,\mathrm{N}$ (b) $10\,\mathrm{m\,s^{-2}}$

3 $81.9\,\mathrm{N},\ 126\,\mathrm{N}$

4 (b) 2.15 p.m., $(10\mathbf{j} + 80\mathbf{i})\,\mathrm{km}$

5 $35\,\mathrm{N},\ 30\,\mathrm{N}$

6 (a)

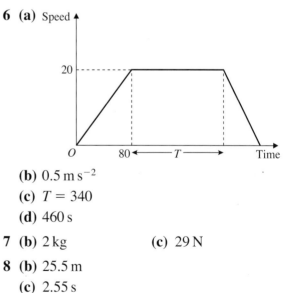

(b) $0.5\,\mathrm{m\,s^{-2}}$
(c) $T = 340$
(d) $460\,\mathrm{s}$

7 (b) $2\,\mathrm{kg}$ (c) $29\,\mathrm{N}$

8 (b) $25.5\,\mathrm{m}$
(c) $2.55\,\mathrm{s}$
(d) $14\,\mathrm{m\,s^{-1}}$

Test yourself answers

Chapter 2

1 **(a)** 50 s   **(b)** $24.2 \, \text{m s}^{-1}$

2 **(a)** 6.2 m   **(b)** 1.6 s   **(c)** Dimensions of ball small, air resistance negligible.

3 **(a)**

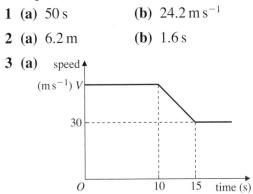

     (b) 42   **(c)** $2.4 \, \text{m s}^{-2}$.

4 **(a)**

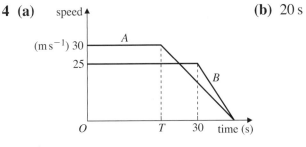

     (b) 20 s   **(c)** $16\frac{2}{3}$ or 16.7

Chapter 3

1 0.25

2 22°

3 **(a)** $4.9 \, \text{m s}^{-2}$ or $\dfrac{g}{2}$ **(b)** $\dfrac{3mg}{2}$

4 **(a)** $4.66 \, \text{m s}^{-2}$ **(b)** $1.57 \, \text{m s}^{-1}$

5 $9 \, \text{m s}^{-1}$

6 **(a)** 4.8 N s   **(b)** 32 N

7 **(a)** $6 \, \text{m s}^{-1}$ **(b)** 45 000 N s   **(c)** $16\frac{2}{3}$ s or 16.7 s

8 **(a)** $2.08 \, \text{m s}^{-1}$ **(b)** 0.651 kN

Chapter 4

1 **(a)** 13 N   **(b)** 25 N

2 7.5 N and 11 N

3 **(a)** 0.98 N   **(b)** 30 N

Chapter 5

1 0.7 m

2 196 N, 19.6 N

3 **(a)** 3   **(b)** 98 N, 49 N

4 **(a)** 2   **(b)** 1.8 m

Chapter 6

1 $|\mathbf{F}_1| = 8.49$ N, $|\mathbf{F}_2| = 6.71$ N

2 **(a)** 86.2 km   **(b)** 320

3 12.30 p.m., $(14\mathbf{i} + 19\mathbf{j})$ km